LOGICAL INVESTIGATIONS

Library of Philosophy and Logic

General Editors:
P. T. Geach, J. L. Mackie, D. H. Mellor, Hilary Putnam,
P. F. Strawson, David Wiggins, Peter Winch

PERCEPTION, EMOTION AND ACTION
IRVING THALBERG

GROUNDLESS BELIEF
MICHAEL WILLIAMS

LOGICAL INVESTIGATIONS

GOTTLOB FREGE

Edited with a Preface by P. T. Geach

Translated by P. T. Geach and R. H. Stoothoff

NEW HAVEN
YALE UNIVERSITY PRESS
1977

This title is published in the United Kingdom
and Commonwealth by Basil Blackwell.

Library of Congress catalog card number: 76–52339

International Standard Book Number: 0–300–02127–5

Printed in Great Britain

Contents

Preface

The three articles here translated were published by Frege in the periodical *Beiträge zur Philosophie des deutschen Idealismus* in 1918 (the first two) and 1923 (the last one). They were intended as chapters of a book, to be called *Logische Untersuchungen*; he never finished it. The title now used is accordingly a translation of Frege's title.

Frege needs no introduction; but readers may be interested in the remarks Wittgenstein made to me about this work in the last months of his life. He took a good deal of interest in the plan Max Black and I had for a little book of Frege translations; and it was through him that I was able to locate some rare works of Frege—the review of Husserl's *Philosophie der Arithmetik* and the essays 'Was ist eine Function?' and 'Die Verneinung'—in the Cambridge University Library. He advised me to translate 'Die Verneinung', but not 'Der Gedanke': that, he considered, was an inferior work—it attacked idealism on its weak side, whereas a worthwhile criticism of idealism would attack it just where it was strongest. Wittgenstein told me he had made this point to Frege in correspondence: Frege could not understand—for him, idealism was the enemy he had long fought, and of course you attack your enemy on his weak side. Wittgenstein never mentioned 'Gedankengefüge' to me, and very likely never knew of its existence.

In spite of Wittgenstein's unfavourable view of 'Der

Gedanke', his own later thought may have been influenced by it. It would not be the only time that Frege's criticism had a delayed action in modifying Wittgenstein's views after he had initially rejected the criticism. For example, Wittgenstein told me how he had reacted to Frege's criticism of the Russellian doctrine of facts—a doctrine still presupposed in the *Tractatus*. By this view, such a fact or complex as knife-to-left-of-book would have the knife and the book as *parts*—though Russell of course avoided the rude four-letter word 'part' and spoke of constituents. Frege asked Wittgenstein if a fact was *bigger* than what it was a fact about; Wittgenstein told me this eventually led him to regard the Russellian view as radically confused, though at the time he thought the criticism silly. Now in 'Der Gedanke' Frege lays down premises from which it is an immediate consequence that certain ideas he plays with in the essay—private sensations with incommunicable qualities, a Cartesian *I* given in an incommunicable way—are really bogus ideas, words with no corresponding thoughts. For Frege affirms (1) that any thought is by its nature communicable, (2) that thoughts about private sensations and sense-qualities, and about the Cartesian *I*, are by their nature incommunicable. It is an immediate consequence that there can be no such thoughts. Frege never drew this conclusion, of course—even though the passage about the two doctors, for whom the patient's pain can be a common object of communicable thoughts without their needing to *have* the pain, comes close to the rejection of pain as a private incommunicable somewhat. But though he never drew this conclusion, Wittgenstein was to draw it.

Of his great debt to Frege Wittgenstein remained conscious to the end of his life. A few days before his death he said to me 'How I wish I could have written like Frege!' And in *Zettel* 712 he wrote 'The style of my sentences is extraordinarily strongly influenced by Frege. And if I wanted to I could establish this influence where at first sight no one would see it.'

P. T. GEACH

University of Leeds
November 1975

Thoughts

Just as 'beautiful' points the way for aesthetics and 'good' for ethics, so do words like 'true' for logic. All sciences have truth as their goal; but logic is also concerned with it in a quite different way: logic has much the same relation to truth as physics has to weight or heat. To discover truths is the task of all sciences; it falls to logic to discern the laws of truth. The word 'law' is used in two senses. When we speak of moral or civil laws we mean prescriptions, which ought to be obeyed but with which actual occurrences are not always in conformity. Laws of nature are general features of what happens in nature, and occurrences in nature are always in accordance with them. It is rather in this sense that I speak of laws of truth. Here of course it is not a matter of what happens but of what is. From the laws of truth there follow prescriptions about asserting, thinking, judging, inferring. And we may very well speak of laws of thought in this way too. But there is at once a danger here of confusing different things. People may very well interpret the expression 'law of thought' by analogy with 'law of nature' and then have in mind general features of thinking as a mental occurrence. A law of thought in this sense would be a psychological law. And so they might come to believe that logic deals with the mental process of thinking and with the psychological laws in accordance with which this takes place. That would be misunderstanding the task of logic, for truth has not here been given its proper place. Error and superstition have causes just as much as correct cognition. Whether what you take for true is false or true, your so taking it comes about in accordance with psychological laws. A derivation from these laws, an explanation of a mental process that ends in taking something to be true, can never take the place of proving what is taken to be true. But may not logical laws also have played a part in this mental process? I do not want to dispute this, but if it is a question of truth this possibility is not enough. For it is also possible that something non-logical played a part in the process and made it swerve from the truth. We can decide only after we have come to know the laws of truth; but then we can probably do without the derivation and explanation of the mental process, if our concern is to decide whether the process terminates in *justifiably* taking something to be true. In order to avoid any misunderstanding and prevent the blurring of the boundary between psychology and logic, I assign to logic the task of discovering the laws of truth, not the laws of taking things to be true or of thinking. The meaning of the word 'true' is spelled out in the laws of truth.

But first I shall attempt to outline roughly how I want to use 'true' in this connexion, so as to exclude irrelevant uses of the word. 'True' is not to be used here in the sense of 'genuine' or 'veracious'; nor yet in the way it sometimes occurs in discussion of artistic questions, when, for example, people speak of truth in art, when truth is set up as the aim of art, when the truth of a work of art or true feeling is spoken of. Again, the word 'true' is prefixed to another word in order to show that the word is to be understood in its proper, unadulterated sense. This use too lies off the path followed here. What I have in mind is that sort of truth which it is the aim of science to discern.

Grammatically, the word 'true' looks like a word for a property. So we want to delimit more closely the region within which truth can be predicated, the region in which there is any question of truth. We find truth predicated of pictures, ideas, sentences, and thoughts. It is striking that visible and audible things turn up here along with things which cannot be perceived with the senses. This suggests that shifts of meaning have taken place. So indeed they have! Is a picture considered as a mere visible and tangible thing really true, and a stone or a leaf not true? Obviously we could not call a picture true unless there were an intention involved. A picture is meant to represent something. (Even an idea is not called true in itself, but only with respect to an intention that the idea should correspond to something.) It might be supposed from this that truth consists in a correspondence of a picture to what it depicts. Now a correspondence is a relation. But this goes against the use of the word 'true', which is not a relative term and contains no indication of anything else to which something is to correspond. If I do not know that a picture is meant to represent Cologne Cathedral then I do not know what to compare the picture with in order to decide on its truth. A correspondence, moreover, can only be perfect if the corresponding things coincide and so are just not different things. It is supposed to be possible to test the genuineness of a banknote by comparing it stereoscopically with a genuine one. But it would be ridiculous to try to compare a gold piece stereoscopically with a twenty-mark note. It would only be possible to compare an idea with a thing if the thing were an idea too. And then, if the first did correspond perfectly with the second, they would coincide. But this is not at all what people intend when they define truth as the correspondence of an idea with something real. For in this case it is essential precisely that the reality shall be distinct from the idea. But then there can be no complete correspondence, no complete truth. So nothing at all would be true; for what is only half true is untrue. Truth does not admit of more and less. —But could we not maintain that there is truth when there is correspondence in a certain respect? But which respect? For in that case what ought we to do so as to decide whether something is true? We should have to inquire whether it is *true* that an idea and a reality, say, correspond in the specified respect. And then we should be confronted by a question of the same kind, and the game could begin again. So the attempted explanation of truth as correspondence breaks down. And any other attempt to define truth also breaks down. For in a definition certain characteristics would have to be specified. And in application to any particular case the question would always arise whether it were *true* that the characteristics were present. So we should be going round in a circle. So it seems likely that the content of the word 'true' is *sui generis* and indefinable.

When we ascribe truth to a picture we do not really mean to ascribe a property which would belong to this picture quite independently of other things; we always have in mind some totally different object and we want to say that the picture corresponds in some way to this object. 'My idea corresponds to Cologne Cathedral' is a sentence, and now it is a matter of the truth of this sentence. So what is improperly called the truth of pictures and ideas is reduced to the truth of sentences. What is it that we call a sentence? A series of sounds, but only if it has a sense (this is not meant to convey that *any* series of sounds that has a sense is a sentence). And when we call a sentence true we really mean that its sense is true. And hence the only thing that raises the question of truth at all is the sense of sentences. Now is the sense of a sentence an idea? In any case, truth does not consist in correspondence of the sense with something else, for otherwise the question of truth would get reiterated to infinity.

Without offering this as a definition, I mean by 'a thought' something for which the question of truth can arise at all. So I count what is false among thoughts no less than what is true[1]. So I can say: thoughts are senses of sentences, without

---

[1] So, similarly, people have said 'a judgement is something which is either true or false'. In fact I use the word 'thought' more or less in the

wishing to assert that the sense of every sentence is a thought. The thought, in itself imperceptible by the senses, gets clothed in the perceptible garb of a sentence, and thereby we are enabled to grasp it. We say a sentence *expresses* a thought.

A thought is something imperceptible: anything the senses can perceive is excluded from the realm of things for which the question of truth arises. Truth is not a quality that answers to a particular kind of sense-impressions. So it is sharply distinguished from the qualities we call by the names 'red', 'bitter', 'lilac-smelling'. But do we not see that the sun has risen? and do we not then also see that this is true? That the sun has risen is not an object emitting rays that reach my eyes; it is not a visible thing like the sun itself. That the sun has risen is recognized to be true on the basis of sense-impressions. But being true is not a sensible, perceptible, property. A thing's being magnetic is also recognized on the basis of sense-impressions of the thing, although this pro-perty does not answer, any more than truth does, to a particular kind of sense-impressions. So far these properties agree. However, we do need sense-impressions in order to recognize a body as magnetic. On the other hand, when I find it is true that I do not smell anything at this moment, I do not do so on the basis of sense-impressions.

All the same it is something worth thinking about that we cannot recognize a property of a thing without at the same time finding the thought *this thing has this property* to be

sense 'judgement' has in the writings of logicians. I hope it will become clear in the sequel why I choose 'thought'. Such an explanation has been objected to on the ground that it makes a division of judgements into true and false judgements—perhaps the least significant of all possible divisions among judgements. But I cannot see that it is a logical fault that a division is given along with the explanation. As for the division's being significant, we shall perhaps find we must hold it in no small esteem, if, as I have said, it is the word 'true' that points the way for logic.

true. So with every property of a thing there is tied up a property of a thought, namely truth. It is also worth noticing that the sentence 'I smell the scent of violets' has just the same content as the sentence 'It is true that I smell the scent of violets'. So it seems, then, that nothing is added to the thought by my ascribing to it the property of truth. And yet is it not a great result when the scientist after much hesitation and laborious researches can finally say 'My conjecture is true'? The meaning of the word 'true' seems to be altogether *sui generis*. May we not be dealing here with something which cannot be called a property in the ordinary sense at all? In spite of this doubt I will begin by expressing myself in accordance with ordinary usage, as if truth were a property, until some more appropriate way of speaking is found.

In order to bring out more precisely what I mean by 'a thought', I shall distinguish various kinds of sentences[2]. We should not wish to deny sense to a command, but this sense is not such that the question of truth could arise for it. Therefore I shall not call the sense of a command a thought. Sentences expressing wishes or requests are ruled out in the same way. Only those sentences in which we communicate or assert something come into the question. But here I do not count exclamations in which one vents one's feelings, groans, sighs, laughs—unless it has been decided by some special convention that they are to communicate something. But how about interrogative sentences? In a word-question* we utter an incomplete sentence, which is meant to be given a true sense just by means of the completion for which we are asking. Word-questions are accordingly left out of considera-

---

[2] I am not using the word 'sentence' here in quite the same sense as grammar does, which also includes subordinate clauses. An isolated subordinate clause does not always have a sense about which the question of truth can arise, whereas the complex sentence to which it belongs has such a sense.

* Frege means a question introduced by an interrogative word like 'who?'

tion here. Propositional questions† are a different matter. We expect to hear 'yes' or 'no'. The answer 'yes' means the same as an assertoric sentence, for in saying 'yes' the speaker presents as true the thought that was already completely contained in the interrogative sentence. This is how a propositional question can be formed from any assertoric sentence. And this is why an exclamation cannot be regarded as a communication: no corresponding propositional question can be formed. An interrogative sentence and an assertoric one contain the same thought; but the assertoric sentence contains something else as well, namely assertion. The interogative sentence contains something more too, namely a request. Therefore two things must be distinguished in an assertoric sentence: the content, which it has in common with the corresponding propositional question; and assertion. The former is the thought or at least contains the thought. So it is possible to express a thought without laying it down as true. The two things are so closely joined in an assertoric sentence that it is easy to overlook their separability. Consequently we distinguish:

(1) the grasp of a thought—thinking,
(2) the acknowledgement of the truth of a thought—the act of judgement[3],
(3) the manifestation of this judgement—assertion.

We have already performed the first act when we form a propositional question. An advance in science usually takes

---

† I.e. yes–no questions: German *Satzfragen*.

[3] It seems to me that thought and judgement have not hitherto been adequately distinguished. Perhaps language is misleading. For we have no particular bit of assertoric sentences which corresponds to assertion; that something is being asserted is implicit rather in the assertoric form. We have the advantage in German that main and subordinate clauses are distinguished by the word-order. However in this connexion we must observe that a subordinate clause may also contain an assertion, and that often neither main nor subordinate clause expresses a complete thought by itself but only the complex sentence does.

B

place in this way: first a thought is grasped, and thus may perhaps be expressed in a propositional question; after appropriate investigations, this thought is finally recognized to be true. We express acknowledgement of truth in the form of an assertoric sentence. We do not need the word 'true' for this. And even when we do use it the properly assertoric force does not lie in it, but in the assertoric sentence-form; and where this form loses its assertoric force the word 'true' cannot put it back again. This happens when we are not speaking seriously. As stage thunder is only sham thunder and a stage fight only a sham fight, so stage assertion is only sham assertion. It is only acting, only fiction. When playing his part the actor is not asserting anything; nor is he lying, even if he says something of whose falsehood he is convinced. In poetry we have the case of thoughts being expressed without being actually put forward as true, in spite of the assertoric form of the sentence; although the poem may suggest to the hearer that he himself should make an assenting judgement. Therefore the question still arises, even about what is presented in the assertoric sentence-form, whether it really contains an assertion. And this question must be answered in the negative if the requisite seriousness is lacking. It is unimportant whether the word 'true' is used here. This explains why it is that nothing seems to be added to a thought by attributing to it the property of truth.

An assertoric sentence often contains, over and above a thought and assertion, a third component not covered by the assertion. This is often meant to act on the feelings and mood of the hearer, or to arouse his imagination. Words like 'regrettably' and 'fortunately' belong here. Such constituents of sentences are more strongly prominent in poetry, but are seldom wholly absent from prose. They occur more rarely in mathematical, physical, or chemical expositions than in historical ones. What are called the humanities are closer to poetry, and are therefore less scientific, than the exact sciences, which are drier in proportion to being more exact; for exact science is directed toward truth and truth alone. Therefore all constituents of sentences not covered by the assertoric force do not belong to scientific exposition; but they are sometimes hard to avoid, even for one who sees the danger connected with them. Where the main thing is to approach by way of intimation what cannot be conceptually grasped, these constituents are fully justified. The more rigorously scientific an exposition is, the less the nationality of its author will be discernible and the easier it will be to translate. On the other hand, the constituents of language to which I here want to call attention make the translation of poetry very difficult, indeed make perfect translation almost always impossible, for it is just in what largely makes the poetic value that languages most differ.

It makes no difference to the thought whether I use the word 'horse' or 'steed' or 'nag' or 'prad'. The assertoric force does not cover the ways in which these words differ. What is called mood, atmosphere, illumination in a poem, what is portrayed by intonation and rhythm, does not belong to the thought.

Much in language serves to aid the hearer's understanding, for instance emphasizing part of a sentence by stress or word-order. Here let us bear in mind words like 'still' and 'already'. Someone using the sentence 'Alfred has still not come' actually says 'Alfred has not come', and at the same time hints—but only hints—that Alfred's arrival is expected. Nobody can say: Since Alfred's arrival is not expected, the sense of the sentence is false. The way that 'but' differs from 'and' is that we use it to intimate that what follows it contrasts with what was to be expected from what preceded it. Such conversational suggestions make no difference to the thought. A sentence can be transformed by changing the verb from active to passive and at the same time making the accusative into the subject. In the same way we may change the dative into the nominative and at the same time replace 'give' by 'receive'. Naturally such transformations are not trivial in every respect; but they do not touch the thought, they do not touch what is true or false. If the inadmissibility of such transformations were recognized as a principle, then any profound logical investigation would be hindered. It is just as important to ignore distinctions that do not touch the heart of the matter, as to make distinctions which concern essentials. But what is essential depends on one's purpose. To a mind concerned with the beauties of language, what is trivial to the logician may seem to be just what is important.

Thus the content of a sentence often goes beyond the thought expressed by it. But the opposite often happens too; the mere wording, which can be made permanent by writing or the gramophone, does not suffice for the expression of the thought. The present tense is used in two ways: first, in order to indicate a time; second, in order to eliminate any temporal restriction, where timelessness or eternity is part of the thought—consider for instance the laws of mathematics. Which of the two cases occurs is not expressed but must be divined. If a time-indication is conveyed by the present tense one must know when the sentence was uttered in order to grasp the thought correctly. Therefore the time of utterance is part of the expression of the thought. If someone wants to say today what he expressed yesterday using the word 'today', he will replace this word with 'yesterday'. Although the thought is the same its verbal expression must be different in order that the change of sense which would otherwise be effected by the differing times of utterance may be cancelled out. The case is the same with words like 'here' and 'there'. In all such cases the mere wording, as it can be preserved in writing, is not the complete expression of the thought; the knowledge of certain conditions accompanying the utterance, which are used as means of expressing the thought, is needed for us to grasp the thought correctly. Pointing the finger, hand gestures, glances may belong here too. The same utterance containing the word 'I' in the mouths of different men will express different thoughts of which some may be true, others false.

The occurrence of the word 'I' in a sentence gives rise to some further questions.

Consider the following case. Dr Gustav Lauben says, 'I was wounded', Leo Peter hears this and remarks some days later, 'Dr Gustav Lauben was wounded'. Does this sentence express the same thought as the one Dr Lauben uttered himself? Suppose that Rudolph Lingens was present when Dr Lauben spoke and now hears what is related by Leo Peter. If the same thought was uttered by Dr Lauben and Leo Peter, then Rudolph Lingens, who is fully master of the language and remembers what Dr Lauben said in his presence, must now know at once from Leo Peter's report that he is speaking of the same thing. But knowledge of the language is a special thing when proper names are involved. It may well be the case that only a few people associate a definite thought with the sentence 'Dr Lauben was wounded'. For complete understanding one needs in this case to know the expression 'Dr Gustav Lauben'. Now if both Leo Peter and Rudolph Lingens mean by 'Dr Gustav Lauben' the doctor who is the only doctor living in a house known to both of them, then they both understand the sentence 'Dr Gustav Lauben was wounded' in the same way; they associate the same thought with it. But it is also possible that Rudolph Lingens does not know Dr Lauben personally and does not know that it was Dr Lauben who recently said 'I was wounded.' In this case Rudolph Lingens cannot know that the same affair is in question. I say, therefore, in this case: the thought which Leo Peter expresses is not the same as that which Dr Lauben uttered.

Suppose further that Herbert Garner knows that Dr

Gustav Lauben was born on 13 September, 1875 in N.N. and this is not true of anyone else; suppose, however, that he does not know where Dr Lauben now lives nor indeed anything else about him. On the other hand, suppose Leo Peter does not know that Dr Lauben was born on 13 September 1875, in N.N. Then as far as the proper name 'Dr Gustav Lauben' is concerned, Herbert Garner and Leo Peter do not speak the same language, although they do in fact refer to the same man with this name; for they do not know that they are doing so. Therefore Herbert Garner does not associate the same thought with the sentence 'Dr Gustav Lauben was wounded' as Leo Peter wants to express with it. To avoid the awkwardness that Herbert Garner and Leo Peter are not speaking the same language, I shall suppose that Leo Peter uses the proper name 'Dr Lauben' and Herbert Garner uses the proper name 'Gustav Lauben'. Then it is possible that Herbert Garner takes the sense of the sentence 'Dr Lauben was wounded' to be true but is misled by false information into taking the sense of the sentence 'Gustav Lauben was wounded' to be false. So given our assumptions these thoughts are different.

Accordingly, with a proper name, it is a matter of the way that the object so designated is presented. This may happen in different ways, and to every such way there corresponds a special sense of a sentence containing the proper name. The different thoughts thus obtained from the same sentences correspond in truth-value, of course; that is to say, if one is true then all are true, and if one is false then all are false. Nevertheless the difference must be recognized. So we must really stipulate that for every proper name there shall be just one associated manner of presentation of the object so designated. It is often unimportant that this stipulation should be fulfilled, but not always.

Now everyone is presented to himself in a special and primitive way, in which he is presented to no-one else. So, when Dr Lauben has the thought that he was wounded, he will probably be basing it on this primitive way in which he is presented to himself. And only Dr Lauben himself can grasp thoughts specified in this way. But now he may want to communicate with others. He cannot communicate a thought he alone can grasp. Therefore, if he now says 'I was wounded', he must use 'I' in a sense which can be grasped by others, perhaps in the sense of 'he who is speaking to you at this moment'; by doing this he makes the conditions accompanying his utterance serve towards the expression of a thought[4].

Yet there is a doubt. Is it at all the same thought which first that man expresses and then this one?

A man who is still unaffected by philosophy first of all gets to know things he can see and touch, can in short perceive with the senses, such as trees, stones and houses, and he is convinced that someone else can equally see and touch the same tree and the same stone as he himself sees and touches. Obviously a thought does not belong with these things. Now can it, nevertheless, like a tree be presented to people as identical?

Even an unphilosophical man soon finds it necessary to recognize an inner world distinct from the outer world, a world of sense-impressions, of creations of his imagination, of sensations, of feelings and moods, a world of inclinations, wishes and decisions. For brevity's sake I want to

---

[4] I am not here in the happy position of a mineralogist who shows his audience a rock-crystal: I cannot put a thought in the hands of my readers with the request that they should examine it from all sides. Something in itself not perceptible by sense, the thought, is presented to the reader—and I must be content with that—wrapped up in a perceptible linguistic form. The pictorial aspect of language presents difficulties. The sensible always breaks in and makes expressions pictorial and so improper. So one fights against language, and I am compelled to occupy myself with language although it is not my proper concern here. I hope I have succeeded in making clear to my readers what I mean by 'a thought'.

use the word 'idea' to cover all these occurrences, except decisions.

Now do thoughts belong to this inner world? Are they ideas? They are obviously not decisions.

How are ideas distinct from the things of the outer world?

First: ideas cannot be seen, or touched, or smelled, or tasted, or heard.

I go for a walk with a companion. I see a green field, I thus have a visual impression of the green. I have it, but I do not see it.

Secondly: ideas are something we have. We have sensations, feelings, moods, inclinations, wishes. An idea that someone has belongs to the content of his consciousness.

The field and the frogs in it, the sun which shines on them, are there no matter whether I look at them or not, but the sense-impression I have of green exists only because of me, I am its owner. It seems absurd to us that a pain, a mood, a wish should go around the world without an owner, independently. A sensation is impossible without a sentient being. The inner world presupposes somebody whose inner world it is.

Thirdly: ideas need an owner. Things of the outer world are on the contrary independent.

My companion and I are convinced that we both see the same field; but each of us has a particular sense-impression of green. I glimpse a strawberry among the green strawberry leaves. My companion cannot find it, he is colour-blind. The colour-impression he gets from the strawberry is not noticeably different from the one he gets from the leaf. Now does my companion see the green leaf as red, or does he see the red berry as green, or does he see both with one colour which I am not acquainted with at all? These are unanswerable, indeed really nonsensical, questions. For when the word 'red' is meant not to state a property of things but to characterize sense-impressions belonging to my consciousness, it is only applicable within the realm of my consciousness. For it is impossible to compare my sense-impression with someone else's. For that, it would be necessary to bring together in one consciousness a sense-impression belonging to one consciousness and a sense-impression belonging to another consciousness. Now even if it were possible to make an idea disappear from one consciousness and at the same time make an idea appear in another consciousness, the question whether it is the same idea would still remain unanswerable. It is so much of the essence of any one of my ideas to be a content of my consciousness, that any idea someone else has is, just as such, different from mine. But might it not be possible that my ideas, the entire content of my consciousness, might be at the same time the content of a more embracing, perhaps Divine consciousness? Only if I were myself part of the Divine Being. But then would they really be my ideas, would I be their owner? This so far oversteps the limits of human understanding that we must leave this possibility out of account. In any case it is impossible for us men to compare other people's ideas with our own. I pick the strawberry, I hold it between my fingers. Now my companion sees it too, this same strawberry; but each of us has his own idea. Nobody else has my idea, but many people can see the same thing. Nobody else has my pain. Someone may have sympathy with me, but still my pain belongs to me and his sympathy to him. He has not got my pain, and I have not got his feeling of sympathy.

Fourthly: every idea has only one owner; no two men have the same idea.

For otherwise it would exist independently of this man and independently of that man. Is that lime-tree my idea? By using the expression 'that lime-tree' in this question I am really already anticipating the answer, for I mean to use this expression to designate what I see and other people too can look at and touch. There are now two possibilities. If my intention is realized, if I do designate something with the expression 'that lime-tree', then the thought expressed in the sentence 'That lime-tree is my idea' must obviously be denied. But if my intention is not realized, if I only think I see without really seeing, if on that account the designation 'that lime-tree' is empty, then I have wandered into the realm of fiction without knowing it or meaning to. In that case neither the content of the sentence 'That lime-tree is my idea' nor the content of the sentence 'That lime-tree is not my idea' is true, for in both cases I have a predication which lacks an object. So then I can refuse to answer the question, on the ground that the content of the sentence 'That lime-tree is my idea' is fictional. I have, of course, got an idea then, but that is not what I am using the words 'that lime-tree' to designate. Now someone might really want to designate one of his ideas with the words 'that lime-tree'. He would then be the owner of that to which he wants to designate with those words, but then he would not see that lime-tree and no-one else would see it or be its owner.

I now return to the question: is a thought an idea? If other people can assent to the thought I express in the Pythagorean theorem just as I do, then it does not belong to the content of my consciousness, I am not its owner; yet I can, nevertheless, acknowledge it as true. However, if what is taken to be the content of the Pythagorean theorem by me and by somebody else is not the same thought at all, we should not really say '*the* Pythagorean theorem', but '*my* Pythagorean theorem', '*his* Pythagorean theorem', and these would be different, for the sense necessarily goes with the sentence. In that case my thought may be the content of my consciousness and his thought the content of his. Could the sense of my Pythagorean theorem be true and the sense of his false? I said that the word 'red' was applicable only in the sphere of my consciousness if it was not meant to state a property of things but to characterize some of my own sense-impressions. Therefore the words 'true' and 'false', as I understand them, might also be applicable only in the realm of my consciousness, if they were not meant to apply to something of which I was not the owner, but to characterize in some way the content of my consciousness. Truth would then be confined to this content and it would remain doubtful whether anything at all similar occurred in the consciousness of others.

If every thought requires an owner and belongs to the contents of his consciousness, then the thought has this owner alone; and there is no science common to many on which many could work, but perhaps I have my science, a totality of thoughts whose owner I am, and another person has his. Each of us is concerned with contents of his own consciousness. No contradiction between the two sciences would then be possible, and it would really be idle to dispute about truth; as idle, indeed almost as ludicrous, as for two people to dispute whether a hundred-mark note were genuine, where each meant the one he himself had in his pocket and understood the word 'genuine' in his own particular sense. If someone takes thoughts to be ideas, what he then accepts as true is, on his own view, the content of consciousness, and does not properly concern other people at all. If he heard from me the opinion that a thought is not an idea he could not dispute it, for, indeed, it would not now concern him.

So the result seems to be: thoughts are neither things in the external world nor ideas.

A third realm must be recognized. Anything belonging to this realm has it in common with ideas that it cannot be perceived by the senses, but has it in common with things that it does not need an owner so as to belong to the contents of his consciousness. Thus for example the thought we have expressed in the Pythagorean theorem is timelessly true, true independently of whether anyone takes it to be true. It needs no owner. It is not true only from the time when it is discovered; just as a planet, even before anyone saw it, was in interaction with other planets.[5]

But I think I hear an odd objection. I have assumed several times that the same thing as I see can also be observed by other people. But what if everything were only a dream? If I only dreamed I was walking in the company of somebody else, if I only dreamed that my companion saw the green field as I did, if it were all only a play performed on the stage of my consciousness, it would be doubtful whether there were things of the external world at all. Perhaps the realm of things is empty and I do not see any things or any men, but only have ideas of which I myself am the owner. An idea, being something which can no more exist independently of me than my feeling of fatigue, cannot be a man, cannot look at the same field together with me, cannot see the strawberry I am holding. It is quite incredible that I really have only my inner world, instead of the whole environment in which I supposed myself to move and to act. And yet this is an inevitable consequence of the thesis that only what is my idea can be the object of my awareness. What would follow from this thesis if it were true? Would there then be other men? It would be possible, but I should know nothing of them. For a man cannot be my idea; consequently, if our thesis were true, he cannot be an object of my awareness either. And so this would undercut any reflections in which I assumed that something was an object for somebody else as it was for myself, for even if this were to happen I should know nothing of it. It would be impossible for me to distinguish something owned by myself from something I did not own. In judging something not to be my idea I would make it into the object of my thinking and, therefore, into

[5] A person sees a thing, has an idea, grasps or thinks a thought. When he grasps or thinks a thought he does not create it but only comes to stand in a certain relation to what already existed—a different relation from seeing a thing or having an idea.

THOUGHTS 19

my idea. On this view, is there a green field? Perhaps, but it would not be visible to me. For if a field is not my idea, it cannot, according to our thesis, be an object of my awareness. But if it is my idea it is invisible, for ideas are not visible. I can indeed have the idea of a green field; but this is not green, for there are no green ideas. Does a missile weighing a hundred kilogrammes exist, according to this view? Perhaps, but I could know nothing of it. If a missile is not my idea then, according to our thesis, it cannot be an object of my awareness, of my thinking. But if a missile were my idea, it would have no weight. I can have an idea of a heavy missile. This then contains the idea of weight as a constituent idea. But this constituent idea is not a property of the whole idea, any more than Germany is a property of Europe. So the consequence is:

Either the thesis that only what is my idea can be the object of my awareness is false, or all my knowledge and perception is limited to the range of my ideas, to the stage of my consciousness. In this case I should have only an inner world and I should know nothing of other people.

It is strange how, in the course of such reflections, opposites turn topsy-turvy. There is, let us suppose, a physiologist of the senses. As is proper for someone investigating nature scientifically, he is at the outset far from supposing the things that he is convinced he sees and touches to be his own ideas. On the contrary, he believes that in sense-impressions he has most reliable evidence of things wholly independent of his feeling, imagining, thinking, which have no need of his consciousness. So little does he consider nerve-fibres and ganglion-cells to be the content of his consciousness that he is on the contrary inclined to regard his consciousness as dependent on nerve-fibres and ganglion-cells. He establishes that light-rays, refracted in the eye, strike the visual nerve-endings and there bring about a change, a stimulus. From this something is transmitted through nerve-fibres to ganglion-cells. Further processes in the nervous system perhaps follow upon this, and colour-impressions arise, and these perhaps combine to make up what we call the idea of a tree. Physical, chemical and physiological occurrences get in between the tree and my idea. Only occurrences in my nervous system are immediately connected with my consciousness—or so it seems—and every observer of the tree has his particular occurrences in his particular nervous system. Now light-rays before they enter my eye, may be reflected by a mirror and diverge as if they came from places behind the mirror. The effects on the visual nerves and all that follows will now take place just as they would if the light-rays had come from a tree behind the mirror and had been propagated undisturbed to the eye. So an idea of a tree will finally occur even though such a tree does not exist at all. The refraction of light too, with the mediation of the eye and nervous system, may give rise to an idea to which nothing at all corresponds. But the stimulation of the visual nerves need not even happen because of light. If lightning strikes near us, we believe we see flames, even though we cannot see the lightning itself. In this case the visual nerve is perhaps stimulated by electric currents occurring in our body as a result of the flash of lightning. If the visual nerve is stimulated by this means in just the way it would be stimulated by light-rays coming from flames, then we believe we see flames. It just depends on the stimulation of the visual nerve, no matter how that itself comes about.

We can go a step further. Properly speaking this stimulation of the visual nerve is not immediately given; it is only an hypothesis. We believe that a thing independent of us stimulates a nerve and by this means produces a sense-impression; but strictly speaking we experience only that end of the process which impinges on our consciousness. Might not this sense-impression, this sensation, which we attribute to a nerve-stimulation, have other causes also, just as the same

nerve-stimulation may arise in different ways? If we call
what happens in our consciousness an idea, then we really
experience only ideas, not their causes. And if the scientist
wants to avoid all mere hypothesis, then he is left just with
ideas; everything dissolves into ideas, even the light-rays,
nerve-fibres and ganglion-cells from which he started. So he
finally undermines the foundations of his own construction. Is everything an idea? Does everything need an owner
without which it could have no existence? I have considered
myself as the owner of my ideas, but am I not myself an
idea? It seems to me as if I were lying in a deck-chair, as if I
could see the toes of a pair of waxed boots, the front part of
a pair of trousers, a waistcoat, buttons, parts of a jacket, in
particular the sleeves, two hands, some hairs of a beard, the
blurred outline of a nose. Am I myself this entire complex of
visual impressions, this aggregate idea? It also seems to me
as if I saw a chair over there. That is an idea. I am not
actually much different from the chair myself, for am I not
myself just a complex of sense-impressions, an idea? But
where then is the owner of these ideas? How do I come to
pick out one of these ideas and set it up as the owner of the
rest? Why need this chosen idea be the idea I like to call 'I'?
Could I not just as well choose the one that I am tempted to
call a chair? Why, after all, have an owner for ideas at all?
An owner would anyhow be something essentially different
from ideas that were just owned; something independent,
not needing any extraneous owner. If everything is idea,
then there is no owner of ideas. And so now once again I
experience opposites turning topsy-turvy. If there is no owner
of ideas then there are also no ideas, for ideas need an owner
and without one they cannot exist. If there is no ruler, there
are also no subjects. The dependence which I found myself
induced to ascribe to the sensation as contrasted with the
sentient being, disappears if there no longer is any owner.
What I called ideas are then independent objects. No reason

remains for granting an exceptional position to that object which I call 'I'.

But is that possible? Can there be an experience without someone to experience it? What would this whole play be without a spectator? Can there be a pain without someone who has it? Being felt necessarily goes with pain, and furthermore someone feeling it necessarily goes with its being felt. But then there *is* something which is not my idea and yet can be the object of my awareness, of my thinking; I myself am such a thing. Or can I be one part of the content of my consciousness, while another part is, perhaps, an idea of the Moon? Does this perhaps take place when I judge that *I* am looking at *the Moon*? Then this first part would have a consciousness, and part of the content of this consciousness would be I myself once more. And so on. Yet it is surely inconceivable that I should be inside myself like this in an infinite nest of boxes, for then there would not be just one I but infinitely many. I am not my own idea; and when I assert something about myself, e.g. that I am not feeling any pain at the moment, then my judgement concerns something which is not a content of my consciousness, is not my idea, namely myself. Therefore that about which I state something is not necessarily my idea. But someone perhaps objects: if I think I have no pain at the moment, does not the word 'I' answer to something in the content of my consciousness? and is that not an idea? That may be so. A certain idea in my consciousness may be associated with the idea of the word 'I'. But then this is one idea among other ideas, and I am its owner as I am the owner of the other ideas. I have an idea of myself, but I am not identical with this idea. What is a content of my consciousness, my idea, should be sharply distinguished from what is an object of my thought. Therefore the thesis that only what belongs to the content of my consciousness can be the object of my awareness, of my thought, is false.

Now the way is clear for me to acknowledge another man likewise as an independent owner of ideas. I have an idea of him, but I do not confuse it with him himself. And if I state something about my brother, I do not state it about the idea that I have of my brother.

The patient who has a pain is the owner of this pain, but the doctor who is treating him and reflects on the cause of this pain is not the owner of the pain. He does not imagine he can relieve the pain by anaesthetizing himself. There may very well be an idea in the doctor's mind that answers to the patient's pain, but that is not the pain, and is not what the doctor is trying to remove. The doctor might consult another doctor. Then one must distinguish: first, the pain, whose owner is the patient; secondly, the first doctor's idea of this pain; thirdly, the second doctor's idea of this pain. This last idea does indeed belong to the content of the second doctor's consciousness, but it is not the object of his reflection; it is rather an aid to reflection, as a drawing may be. The two doctors have as their common object of thought the patient's pain, which they do not own. It may be seen from this that not only a thing but also an idea may be a common object of thought for people who do not have the idea.

In this way, it seems to me, the matter becomes intelligible. If man could not think and could not take as the object of his thought something of which he was not the owner, he would have an inner world but no environment. But may this not be based on a mistake? I am convinced that the idea I associate with the words 'my brother' corresponds to something that is not my idea and about which I can say something. But may I not be making a mistake about this? Such mistakes do happen. We then, against our will, lapse into fiction. Yes, indeed! By the step with which I win an environment for myself I expose myself to the risk of error. And here I come up against a further difference between my inner world and the external world. I cannot doubt that I have a

C

visual impression of green, but it is not so certain that I see a lime-leaf. So, contrary to widespread views, we find certainty in the inner world, while doubt never altogether leaves us in our excursions into the external world. But the probability is nevertheless in many cases hard to distinguish from certainty, so we can venture to judge about things in the external world. And we must make this venture even at the risk of error if we do not want to fall into far greater dangers.

As the result of these last considerations I lay down the following: not everything that can be the object of my acquaintance is an idea. I, being owner of ideas, am not myself an idea. Nothing now stops me from acknowledging other men to be owners of ideas, just as I am myself. And, once given the possibility, the probability is very great, so great that it is in my opinion no longer distinguishable from certainty. Would there be a science of history otherwise? Would not all moral theory, all law, otherwise collapse? What would be left of religion? The natural sciences too could only be assessed as fables like astrology and alchemy. Thus the reflections I have set forth on the assumption that there are other men besides myself, who can make the same thing the object of their consideration, their thinking, remain in force without any essential weakening.

Not everything is an idea. Thus I can also acknowledge thoughts as independent of me; other men can grasp them just as much as I; I can acknowledge a science in which many can be engaged in research. We are not owners of thoughts as we are owners of our ideas. We do not *have* a thought as we have, say, a sense-impression, but we also do not *see* a thought as we see, say, a star. So it is advisable to choose a special expression; the word 'grasp' suggests itself for the purpose[6]. To the grasping of thoughts there must

---

[6] The expression 'grasp' is as metaphorical as 'content of consciousness'. The nature of language does not permit anything else. What I hold in my hand can certainly be regarded as the content of my hand; but all

then correspond a special mental capacity, the power of thinking.

In thinking we do not produce thoughts, we grasp them. For what I have called thoughts stand in the closest connexion with truth. What I acknowledge as true, I judge to be true quite apart from my acknowledging its truth or even thinking about it. That someone thinks it has nothing to do with the truth of a thought. 'Facts, facts, facts' cries the scientist if he wants to bring home the necessity of a firm foundation for science. What is a fact? A fact is a thought that is true. But the scientist will surely not acknowledge something to be the firm foundation of science if it depends on men's varying states of consciousness. The work of science does not consist in creation, but in the discovery of true thoughts. The astronomer can apply a mathematical truth in the investigation of long past events which took place when—on Earth at least— no-one had yet recognized that truth. He can do this because the truth of a thought is timeless. Therefore that truth cannot have come to be only upon its discovery.

Not everything is an idea. Otherwise psychology would contain all the sciences within it, or at least it would be the supreme judge over all the sciences. Otherwise psychology would rule even over logic and mathematics. But nothing would be a greater misunderstanding of mathematics than making it subordinate to psychology. Neither logic nor mathematics has the task of investigating minds and contents of consciousness owned by individual men. Their task could perhaps be represented rather as the investigation of *the* mind; of *the* mind, not of minds.

The grasp of a thought presupposes someone who grasps it, who thinks. He is the owner of the thinking, not of the

the same it is the content of my hand in quite another and a more extraneous way than are the bones and muscles of which the hand consists or again the tensions these undergo.

thought. Although the thought does not belong with the contents of the thinker's consciousness, there must be something in his consciousness that is aimed at the thought. But this should not be confused with the thought itself. Similarly Algol itself is different from the idea someone has of Algol.

A thought belongs neither to my inner world as an idea, nor yet to the external world, the world of things perceptible by the senses.

This consequence, however cogently it may follow from the exposition, will nevertheless perhaps not be accepted without opposition. It will, I think, seem impossible to some people to obtain information about something not belonging to the inner world except by sense-perception. Sense-perception indeed is often thought to be the most certain, even the sole, source of knowledge about everything that does not belong to the inner world. But with what right? For sense-perception has as necessary constituents our sense-impressions and these are a part of the inner world. In any case two men do not have the same sense-impressions though they may have similar ones. Sense-impressions alone do not reveal the external world to us. Perhaps there is a being that has only sense-impressions without seeing or touching things. To have visual impressions is not to see things. How does it happen that I see the tree just there where I do see it? Obviously it depends on the visual impressions I have and on the particular sort which occur because I see with two eyes. On each of the two retinas there arises, physically speaking, a particular image. Someone else sees the tree in the same place. He also has two retinal images but they differ from mine. We must assume that these retinal images determine our impressions. Consequently the visual impressions we have are not only not the same, but markedly different from each other. And yet we move about in the same external world. Having visual impressions is certainly necessary for seeing things, but not sufficient. What must still be added is not anything sensible. And yet this is just what opens up the external world for us; for without this non-sensible something everyone would remain shut up in his inner world. So perhaps, since the decisive factor lies in the non-sensible, something non-sensible, even without the co-operation of sense-impressions, could also lead us out of the inner world and enable us to grasp thoughts. Outside our inner world we should have to distinguish the external world proper of sensible, perceptible things and the realm of what is non-sensibly perceptible. We should need something non-sensible for the recognition of both realms; but for the sense-perception of things we should need sense-impressions as well, and these belong entirely to the inner world. So the distinction between the ways in which a thing and a thought are given mainly consists in something which is assignable, not to either of the two realms, but to the inner world. Thus I cannot find this distinction to be so great as to make impossible the presentation of a thought that does not belong to the inner world.

A thought, admittedly, is not the sort of thing to which it is usual to apply the term 'actual'. The world of actuality is a world in which this acts on that and changes it and again undergoes reactions itself and is changed by them. All this is a process in time. We will hardly admit what is timeless and unchangeable to be actual. Now is a thought changeable or is it timeless? The thought we express by the Pythagorean theorem is surely timeless, eternal, unvarying. But are there not thoughts which are true today but false in six months' time? The thought, for example, that the tree there is covered with green leaves, will surely be false in six months' time. No, for it is not the same thought at all. The words 'This tree is covered with green leaves' are not sufficient by themselves to constitute the expression of thought, for the time of utterance is involved as well. Without the time-specification thus given we have not a complete thought, i.e. we have no thought at all. Only a sentence with the time-specification filled out, a sentence complete in every respect, expresses a thought. But this thought, if it is true, is true not only today or tomorrow but timelessly. Thus the present tense in 'is true' does not refer to the speaker's present; it is, if the expression be permitted, a tense of timelessness. If we merely use the assertoric sentence-form and avoid the word 'true', two things must be distinguished, the expression of the thought and assertion. The time-specification that may be contained in the sentence belongs only to the expression of the thought; the truth, which we acknowledge by using the assertoric sentence-form, is timeless. To be sure the same words, on account of the variability of language with time, may take on another sense, express another thought; this change, however, relates only to the linguistic realm.

And yet what value could there be for us in the eternally unchangeable, which could neither be acted upon nor act on us? Something entirely and in every respect inactive would be quite unactual, and so far as we are concerned it would not be there. Even the timeless, if it is to be anything for us, must somehow be implicated with the temporal. What would a thought be for me if it were never grasped by me? But by grasping a thought I come into a relation to it, and it to me. It is possible that the same thought as is thought by me today was not thought by me yesterday. Of course this does away with strict timelessness. But we may be inclined to distinguish between essential and inessential properties and to regard something as timeless if the changes it undergoes involve only inessential properties. A property of a thought will be called inessential if it consists in, or follows from, the fact that this thought is grasped by a thinker.

How does a thought act? By being grasped and taken to be true. This is a process in the inner world of a thinker which may have further consequences in this inner world, and which may also encroach on the sphere of the will and make itself noticeable in the outer world as well. If, for example, I grasp the thought we express by the theorem of Pythagoras, the consequence may be that I recognize it to be true, and further that I apply it in making a decision, which brings about the acceleration of masses. This is how our actions are usually led up to by acts of thinking and judging. And so thought may indirectly influence the motion of masses. The influence of man on man is brought about for the most part by thoughts. People communicate thoughts. How do they do this? They bring about changes in the common external world, and these are meant to be perceived by someone else, and so give him a chance to grasp a thought and take it to be true. Could the great events of world history have come about without the communication of thoughts? And yet we are inclined to regard thoughts as unactual, because they appear to do nothing in relation to events, whereas thinking, judging, stating, understanding, in general doing things, are affairs that concern men. How very different the actuality of a hammer appears, compared with that of a thought! How different a process handing over a hammer is from communicating a thought! The hammer passes from one control to another, it is gripped, it undergoes pressure, and thus its density, the disposition of its parts, is locally changed. There is nothing of all this with a thought. It does not leave the control of the communicator by being communicated, for after all man has no power over it. When a thought is grasped, it at first only brings about changes in the inner world of the one who grasps it; yet it remains untouched in the core of its essence, for the changes it undergoes affect only inessential properties. There is lacking here something we observe everywhere in physical process—reciprocal action. Thoughts are not wholly unactual but their actuality is quite different from the actuality of things. And their action is brought about by a performance of the thinker; without this they would be inactive, at least as far as we can see. And yet the thinker does not create them but must take them as they are. They can be true without being grasped by a thinker; and they are not wholly unactual even then, at least if they *could* be grasped and so brought into action.

Negation[†]

A propositional question contains a demand that we should either acknowledge the truth of a thought, or reject it as false. In order that we may meet this demand correctly, two things are requisite: first, the wording of the question must enable us to recognize without any doubt the thought that is referred to; secondly, this thought must not belong to fiction. I always assume in what follows that these conditions are fulfilled. The answer to a question[7] is an assertion based upon a judgement; this is so equally whether the answer is affirmative or negative.

Here, however, a difficulty arises. If a thought has being by being true, then the expression 'false thought' is just as self-contradictory, as 'thought that has no being.' In that case the expression 'the thought: three is greater than five' is an empty one; and accordingly in science it must not be used at all—except between quotation-marks. In that case we may not say 'that three is greater than five is false'; for the grammatical subject is empty.

But can we not at least ask if something is true? In a question we can distinguish between the demand for a decision and the special content of the question, the point we are to decide. In what follows I shall call this special content simply the content of the question, or the sense of the corresponding interrogative sentence. Now has the interrogative sentence

† First published in *Beiträge zur Philosophie des deutschen Idealismus*, vol. 1 (1919); pp. 143–57.

7 Here and in what follows I always mean a propositional question when I just write 'question.'

'Is 3 greater than 5?'

a sense, if the being of a thought consists in its being true? If not, the question cannot have a thought as its content; and one is inclined to say that the interrogative sentence has no sense at all. But this surely comes about because we see the falsity at once. Has the interrogative sentence

'Is $(21/20)^{100}$ greater than $\sqrt[10]{10^{21}}$?'

got a sense? If we had worked out that the answer must be affirmative, we could accept the interrogative sentence as making sense, for it would have a thought as its sense. But what if the answer had to be negative? In that case, on our supposition, we should have no thought that was the sense of the question. But surely the interrogative sentence must have some sense or other, if it is to contain a question at all. And are we really not asking for something in this sentence? May we not be wanting to get an answer to it? In that case, it depends on the answer whether we are to suppose that the question has a thought as its content. But it must be already possible to grasp the sense of the interrogative sentence before answering the question; for otherwise no answer would be possible at all. So that which we can grasp as the sense of the interrogative sentence before answering the question—and only this can properly be called the sense of the interrogative sentence—cannot be a thought, if the being of a thought consists in being true. 'But is it not a truth that the Sun is bigger than the Moon? And does not the being of a truth just consist in its being true? Must we not there- fore recognize after all that the sense of the interrogative sentence:

'Is the Sun bigger than the Moon?'

is a truth, a thought whose being consists in its being true?' No! Truth cannot go along with the sense of an interrogative sentence; that would contradict the very nature of a question.

The content of a question is that as to which we must judge. Consequently truth cannot be counted as going along with the content of the question. When I raise the question whether the Sun is bigger than the Moon, I am seeing the sense of the interrogative sentence

'Is the Sun bigger than the Moon?'

Now if this sense were a thought whose being consisted in its being true, then I should at the same time see that this sense was true. Grasping the sense would at the same time be an act of judging; and the utterance of the interrogative sentence would at the same time be an assertion, and so an answer to the question. But in an interrogative sentence neither the truth nor the falsity of the sense may be asserted. Hence an interrogative sentence has not as its sense something whose being consists in its being true. The very nature of a question demands a separation between the acts of grasping a sense and of judging. And since the sense of an interrogative sentence is always also inherent in the assertoric sentence that gives an answer to the question, this separation must be carried out for assertoric sentences too. It is a matter of what we take the word 'thought' to mean. In any case, we need a short term for what can be the sense of an interrogative sentence. I call this a thought. If we use language this way, not all thoughts are true. The being of a thought thus does not consist in its being true. We must recognize that there are thoughts in this sense, since we use questions in scientific work; for the investigator must sometimes content himself with raising a question, until he is able to answer it. In raising the question he is grasping a thought. Thus I may also say: The investigator must sometimes content himself with grasping a thought. This is anyhow already a step towards the goal, even if it is not yet a judgement. There must, then, be thoughts, in the sense I have assigned to the word. Thoughts that perhaps turn out later on to be false have a justifiable use in science, and must not be treated as having no being. Consider indirect proof; here knowledge of the truth is attained precisely through our grasping a false thought. The teacher says 'Suppose a were not equal to b.' A beginner at once thinks 'What nonsense! I can see that a *is* equal to b'; he is confusing the senselessness of a sentence with the falsity of the thought expressed in it.

Of course we cannot infer anything from a false thought; but the false thought may be part of a true thought, from which something can be inferred. The thought contained in the sentence:

'If the accused was in Rome at the time of the deed, he did not commit the murder'[8]

may be acknowledged to be true by someone who does not know if the accused was in Rome at the time of the deed nor if he committed the murder. Of the two component thoughts contained in the whole, neither the antecedent nor the consequent is being uttered assertively when the whole is presented as true. We have then only a single act of judgement, but three thoughts, viz. the whole thought, the antecedent, and the consequent. If one of the clauses were senseless, the whole would be senseless. From this we see what a difference it makes whether a sentence is senseless or on the contrary expresses a false thought. Now for thoughts consisting of an antecedent and a consequent there obtains the law that, without prejudice to the truth, the opposite of the antecedent may become the consequent, and the opposite of the consequent the antecedent. The English call this procedure *contraposition*.

According to this law, we may pass from the proposition

[8] Here we must suppose that these words by themselves do not contain the thought in its entirety; that we must gather from the circumstances in which they are uttered how to supplement them so as to get a complete thought.

'If $(21/20)^{100}$ is greater than $\sqrt[100]{10^{21}}$, then $(21/20)^{1000}$ is greater than 10^{21}'

to the proposition

'If $(21/20)^{1000}$ is not greater than 10^{21}, then $(21/20)^{100}$ is not greater than $\sqrt[10]{10^{21}}$'

And such transitions are important for indirect proofs, which would otherwise not be possible.

Now if the first complex thought has a true antecedent, viz, $(21/20)^{100}$ *is greater than* $\sqrt[10]{10^{21}}$, then the second complex thought has a false consequent, viz. $(21/20)^{100}$ *is not greater than* $\sqrt[10]{10^{21}}$. So anybody that admits the legitimacy of our transition from *modus ponens* to *modus tollens* must acknowledge that even a false thought has being; for otherwise either only the consequent would be left in the *modus ponens* or only the antecedent in the *modus tollens*; and one of these would likewise be abolished as a nonentity.

The being of a thought may also be taken to lie in the possibility of different thinkers' grasping the thought as one and the same thought. In that case the fact that a thought had no being would consist in several thinkers' each associating with the sentence a sense of his own; this sense would in that case be a content of his particular consciousness, so that there would be no *common* sense that could be grasped by several people. Now is a false thought a thought that in this sense has no being? In that case investigators who had discussed among themselves whether bovine tuberculosis is communicable to men, and had finally agreed that such communicability did not exist, would be in the same position as people who had used in conversation the expression 'this rainbow,' and now came to see that they had not been designating anything by these words, since what each of them had had was a phenomenon of which he himself was the owner. The investigators would have to realize that they had been deceived by a false appearance; for the presupposition that could alone have made all their activity and talk reasonable would have turned out not to be fulfilled; they would not have been giving the question that they discussed a sense common to all of them.

But it must be possible to put a question to which the true answer is negative. The content of such a question is, in my terminology, a thought. It must be possible for several people who hear the same interrogative sentence to grasp the same sense and recognize the falsity of it. Trial by jury would assuredly be a silly arrangement if it could not be assumed that each of the jurymen could understand the question at issue in the same sense. So the sense of an interrogative sentence, even when the question has to be answered in the negative, is something that can be grasped by several people.

What else would follow if the truth of a thought consisted in the possibility of its being grasped by several people as one and the same thing, whereas a sentence that expressed something false had no sense common to several people?

If a thought is true and is a complex of thoughts of which one is false, then the whole thought could be grasped by several people as one and the same thing, but the false component thought could not. Such a case may occur. E.g. it may be that the following assertion is justifiably made before a jury: 'If the accused was in Rome at the time of the deed, he did not commit the murder'; and it may be false that the accused was in Rome at the time of the deed. In that case the jurymen could grasp the same thought when they heard the sentence 'If the accused was in Rome at the time of the deed, he did not commit the murder,' whereas each of them would associate a sense of his own with the *if*-clause. Is this possible? Can a thought that is present to all the jurymen as one and the same thing have a part that is not common to all of them? If the whole needs no owner, no part of it needs an owner.

So a false thought is not a thought that has no being—not even if we take 'being' to mean 'not needing an owner.' A false thought must be admitted, not indeed as true, but as sometimes indispensable: first, as the sense of an interrogative sentence; secondly, as part of a hypothetical thought-complex; thirdly, in negation. It must be possible to negate a false thought, and for this I need the thought; I cannot negate what is not there. And by negation I cannot transform something that needs me as its owner into something of which I am not the owner, and which can be grasped by several people as one and the same thing.

Now is negation of a thought to be regarded as dissolution of the thought into its component parts? By their negative verdict the jury can in no way alter the make-up of the thought that the question presented to them expresses. The thought is true or false quite independently of their giving a right or a wrong verdict in regard to it. And if it is false it is still a thought. If after the jury's verdict there is no thought at all, but only fragments of thought, then the same was already the case before the verdict; in what looked like a question, the jury were not presented with any thought at all, but only with fragments of thought; they had nothing to pass a verdict on.

Our act of judgement can in no way alter the make-up of a thought. We can only acknowledge what is there. A true thought cannot be affected by our act of judgement. In the sentence that expresses the thought we can insert a 'not'; and the sentence we thus get does not contain a non-thought (as I have shown) but may be quite justifiably used in antecedent or consequent in a hypothetical sentence complex. Only, since it is false, it may not be uttered assertively. But this procedure does not touch the original thought in any way; it remains true as before.

Can we affect a false thought somehow by negating it? We cannot do this either; for a false thought is still a thought and may occur as a component part of a true thought. The sentence

'3 is greater than 5,'

uttered non-assertively, has a false sense; if we insert a 'not,' we get

'3 is not greater than 5,'

a sentence that may be uttered assertively. There is no trace here of a dissolution of the thought, a separation of its parts.

How, indeed, could a thought be dissolved? How could the interconnexion of its parts be split up? The world of thoughts has a model in the world of sentences, expressions, words, signs. To the structure of the thought there corresponds the compounding of words into a sentence; and here the order is in general not indifferent. To the dissolution or destruction of the thought there must accordingly correspond a tearing apart of the words, such as happens, e.g., if a sentence written on paper is cut up with scissors, so that on each scrap of paper there stands the expression for part of a thought. These scraps can then be shuffled at will or carried away by the wind; the connexion is dissolved, the original order can no longer be recognized. Is this what happens when we negate a thought? No! The thought would undoubtedly survive even this execution of it in effigy. What we do is to insert the word 'not,' and, apart from this, leave the word-order unaltered. The original wording can still be recognized; the order may not be altered at will. Is this dissolution, separation? Quite the reverse! it results in a firmly-built structure.

Consideration of the law *duplex negatio affirmat* makes it specially plain to see that negation has no separating or dissolving effect. I start with the sentence

'The Schneekoppe is higher than the Brocken.'

By putting in a 'not' I get:

'The Schneekoppe is not higher than the Brocken.'

(Both sentences are supposed to be uttered non-assertively.) A second negation would produce something like the sentence

'It is not true that the Schneekoppe is not higher than the Brocken.'

We already know that the first negation cannot effect any dissolution of the thought; but all the same let us suppose for once that after the first negation we had only fragments of a thought. We should then have to suppose that the second negation could put these fragments together again. Negation would thus be like a sword that could heal on again the limbs it had cut off. But here the greatest care would be wanted. The parts of the thought have lost all connexion and inter-relation on account of its being negated the first time. So by carelessly employing the healing power of negation, we might easily get the sentence:

'The Brocken is higher than the Schneekoppe.'

No non-thought is turned into a thought by negation, just as no thought is turned into a non-thought by negation.

A sentence with the word 'not' in its predicate may, like any other, express a thought that can be made into the content of a question; and this, like any propositional question, leaves open our decision as to the answer.

What then are these objects, which negation is supposed to separate? Not parts of sentences; equally, not parts of a thought. Things in the outside world? They do not bother about our negating. Mental images in the interior world of the person who negates? But then how does the juryman know which of his images he ought to separate in given

D

circumstances? The question put before him does not indicate any to him. It may evoke images in him. But the images
evoked in the jurymen's inner worlds are different; and in
that case each juryman would perform his own act of separation in his own inner world, and this would not be a
verdict.

It thus appears impossible to state what really is dissolved,
split up, or separated by the act of negation.

With the belief that negation has a dissolving or separating power there hangs together the view that a negative
thought is less useful than an affirmative one. But still it cannot be regarded as wholly useless. Consider the inference:

'If the accused was not in Berlin at the time of the murder, he did not commit the murder; now the accused was
not in Berlin at the time of the murder; therefore he did
not commit the murder,'

and compare it with the inference:

'If the accused was in Rome at the time of the murder,
he did not commit the murder; now the accused was in
Rome at the time of the murder; therefore he did not
commit the murder.'

Both inferences proceed in the same form, and there is not
the least ground in the nature of the case for our distinguish-
ing between negative and affirmative premises when we are
expressing the law of inference here involved. People speak
of affirmative and negative judgements; even Kant does so.
Translated into my terminology, this would be a distinction
between affirmative and negative thoughts. For logic at any
rate such a distinction is wholly unnecessary; its ground
must be sought outside logic. I know of no logical principle
whose verbal expression makes it necessary, or even prefer-

able, to use these terms.[9] In any science in which it is a question of conformity to laws, the thing that we must always ask is: What technical expressions are necessary or at least useful, in order to give precise expression to the laws of this science? What does not stand this test cometh of evil.[A]

What is more, it is by no means easy to state what is a negative judgement (thought). Consider the sentences 'Christ is immortal,' 'Christ lives for ever,' 'Christ is not immortal,' 'Christ is mortal,' 'Christ does not live for ever.' Now which of the thoughts we have here is affirmative, which negative?

We usually suppose that negation extends to the whole thought when 'not' is attached to the verb of the predicate. But sometimes the negative word grammatically forms part of the subject, as in the sentence 'no man lives to be more than a hundred.' A negation may occur anywhere in a sentence without making the thought indubitably negative. We see what tricky questions the expression 'negative judgement (thought)' my lead to. The result may be endless disputes, carried on with the greatest subtlety, and nevertheless essentially sterile. Accordingly I am in favour of dropping the distinction between negative and affirmative judgements or thoughts until such time as we have a criterion enabling us to distinguish with certainty in any given case between a negative and an affirmative judgement. When we have such a criterion we shall also see what benefit may be expected from this distinction. For the present I still doubt whether this will be achieved. The criterion cannot be derived from language; for languages are unreliable on logical questions.

---

[9] Accordingly, in my essay *Thoughts* (*Beiträge zur Philosophie des deutschen Idealismus*, Vol. i, p. 58) I likewise made no use of the expression 'negative thought.' The distinction between negative and affirmative thoughts would only have confused the matter. At no point would there have been occasion to assert something about affirmative thoughts, excluding negative ones, or to assert something about negative thoughts, excluding affirmative ones.

[A] An apparent allusion to Matthew v. 37!

It is indeed not the least of the logician's tasks to indicate the pitfalls laid by language in the way of the thinker.

After refuting errors, it may be useful to trace the sources from which they have flowed. One source, I think, in this case is the desire to give definitions of the concepts one means to employ. It is certainly praiseworthy to try to make clear to oneself as far as possible the sense one associates with a word. But here we must not forget that not everything can be defined. If we insist at any price on defining what is essentially indefinable, we readily fasten upon inessential accessories, and thus start the inquiry on a wrong track at the very outset. And this is certainly what has happened to many people, who have tried to explain what a judgement is and so have hit upon compositeness.[10] The judgement is composed of parts that have a certain order, an interconnexion, stand in mutual relations; but for what whole we do not get this?

There is another mistake associated with this one: viz. the

---

[10] We are probably best in accord with ordinary usage if we take a judgement to be an act of judging, as a leap is an act of leaping. Of course this leaves the kernel of the difficulty uncracked; it now lies in the word 'judging'. Judging, we may say, is acknowledging the truth of something; what is acknowledged to be true can only be thought. The original kernel now seems to have cracked in two; one part of it lies in the word 'thought' and the other in the word 'true'. Here, for sure, we must stop. The impossibility of an infinite regress in definition is something we must be prepared for in advance.

If a judgement is an act, it happens at a certain time and thereafter belongs to the past. With an act there also belongs an agent, and we do not know the act completely if we do not know the agent. In that case, we cannot speak of a synthetic judgement in the usual sense. If we call it a synthetic judgement that through two points only one straight line passes, then we are taking 'judgement' to mean, not an act performed by a definite man at a definite time, but something timelessly true, even if its being true is not acknowledged by any human being. If we call this sort of thing a truth, then we may perhaps with advantage say 'synthetic truth' instead of 'synthetic judgement.' If we do nevertheless prefer the expression 'synthetic judgement,' we must leave out of consideration the sense of the verb 'to judge.'

view that the judging subject sets up the connexion or order of the parts in the act of judging and thereby brings the judgement into existence. Here the act of grasping a thought and the acknowledgement of its truth are not kept separate. In many cases, of course, one of these acts follows so directly upon the other that they seem to fuse into one act; but not so in all cases. Years of laborious investigations may come between grasping a thought and acknowledging its truth. It is obvious that here the act of judging did not make the thought or set its parts in order; for the thought was already there. But even the act of grasping a thought is not a production of the thought, is not an act of setting its parts in order; for the thought was already true, and so was already there with its parts in order, before it was grasped. A traveller who crosses a mountain-range does not thereby make the mountain-range; no more does the judging subject make a thought by acknowledging its truth. If he did, the same thought could not be acknowledged as true by one man yesterday and another man to-day; indeed, the same man could not recognize the same thought as true at different times—unless we supposed that the existence of the thought was an intermittent one.

If someone thinks it within his power to produce by an act of judgement that which, in judging, he acknowledges to be true, by setting up an interconnexion, an order, among its parts; then it is easy for him to credit himself also with the power of destroying it. As destruction is opposed to construction, to setting up order and interconnexion, so also negating seems to be opposed to judging; and people easily come to suppose that the interconnexion is broken up by the act of negation just as it is built up by the act of judgement. Thus judging and negating look like a pair of polar opposites, which, being a pair, are co-ordinate; a pair comparable, e.g., to oxidation and reduction in chemistry. But when once we see that no interconnexion is set up by our judging;

that the parts of the thought were already in their order before our judging; then everything appears in a different light. It must be pointed out yet once more that to grasp a thought is not yet to judge; that we may express a thought in a sentence without asserting its truth; that a negative word may be contained in the predicate of a sentence, in which case the sense of this word is part of the sense of the sentence, part of the thought; that by inserting a 'not' in the predicate of a sentence meant to be uttered non-assertively, we get a sentence that expresses a thought, as the original one did. If we call such a transition, from a thought to its opposite, negating the thought, then negating in this sense is not co-ordinate with judging, and may not be regarded as the polar opposite of judging; for what matters in judging is always the truth, whereas we may pass from a thought to its opposite without asking which is true. To exclude misunderstanding, let it be further observed that this transition occurs in the consciousness of a thinker, whereas the thoughts that are the *termini a quo* and *ad quem* of the transition were already in being before it occurred; so that this psychical event makes no difference to the make-up and the mutual relations of the thoughts.

Perhaps the act of negating, which maintains a questionable existence as the polar opposite of judging, is a chimerical construction, formed by a fusion of the act of judging with the negation that I have acknowledged as a possible component of a thought, and to which there corresponds in language the word 'not' as part of the predicate—a chimerical construction, because these parts are quite different in kind. The act of judging is a psychical process, and as such it needs a judging subject as its owner; negation on the other hand is part of a thought, and as such, like the thought itself, it needs no owner, must not be regarded as a content of a consciousness. And yet it is not quite incomprehensible how there can arise at least the illusion of such a chimerical construction. Language has no special word or syllable to express assertion; assertive force is supplied by the form of the assertoric sentence, which is specially well-marked in the predicate. On the other hand the word 'not' stands in intimate connexion with the predicate and may be regarded as part of it. Thus a connexion may seem to be formed between the word 'not' and the assertoric force in language that answers to the act of judging.

But it is a nuisance to distinguish between the two ways of negating. Really my only aim in introducing the polar opposite of judging was to accommodate myself to a way of thinking that is foreign to me. I now return to my previous way of speaking. What I have just been designating as the polar opposite of judging I will now regard as a second way of judging—without thereby admitting that there is such a second way. I shall thus be comprising both polar opposites under the common term 'judging'; this may be done, for polar opposites certainly do belong together. The question will then have to be put as follows:

Are there two different ways of judging, of which one is used for the affirmative, and the other for the negative, answer to a question? Or is judging the same act in both cases? Does negating go along with judging? Or is negation part of the thought that underlies the act of judging? Does judging consist, even in the case of a negative answer to a question, in acknowledging the truth of a thought? In that case the thought will not be the one directly contained in the question, but the opposite of this.

Let the question run, e.g., as follows: 'Did the accused intentionally set fire to his house?' How can the answer take the form of an assertoric sentence, if it turns out to be negative? If there is a special way of judging for when we deny, we must correspondingly have a special form of assertion. I may, e.g., say in this case 'it is false that . . .' and lay it down that this must always have assertoric force attached to it. Thus the answer will run something like this: 'It is false that the accused intentionally set fire to his house.' If on the other hand there is only one way of judging, we shall say assertorically: 'The accused did not intentionally set fire to his house.' And here we shall be presenting as something true the opposite thought to the one expressed in the question. The word 'not' here belongs with the expression of this thought. I now refer back to the two inferences I compared together just now. The second premise of the first inference was the negative answer to the question 'was the accused in Berlin at the time of the murder?'—in fact, the answer that we fixed upon in case there is only one way of judging. The thought contained in this premise is contained in the *if*-clause of the first premise, but there it is uttered non-assertively. The second premise of the second inference was the affirmative answer to the question 'Was the accused in Rome at the time of the murder?' These inferences proceed on the same principle, which is in good agreement with the view that judging is the same act whether the answer to a question is affirmative or negative. If on the other hand we had to recognize a special way of judging for the negative case—and correspondingly, in the realm of words and sentences, a special form of assertion—the matter would be otherwise. The first premise of the first inference would run as before:

'If the accused was not in Berlin at the time of the murder, he did not commit the murder.'

Here we could not say 'If it is false that the accused was in Berlin at the time of the murder'; for we have laid it down that to the words 'it is false that' assertoric force must always be attached; but in acknowledging the truth of this first premise we are not acknowledging the truth either of its antecedent or of its consequent. The second premise on

the other hand must now run: 'It is false that the accused
was in Berlin at the time of the murder'; for being a premise
it must be uttered assertively. The inference now cannot be
performed in the same way as before; for the thought in the
second premise no longer coincides with the antecedent of
the first premise; it is now the thought that the accused *was*
in Berlin at the time of the murder. If nevertheless we want
to allow that the inference is valid, we are thereby acknowledging that the second premise contains the thought that the
accused was *not* in Berlin at the time of the murder. This
involves separating negation from the act of judging, extract-
ing it from the sense of 'it is false that...', and uniting
negation with the thought.

Thus the assumption of two different ways of judging
must be rejected. But what hangs on this decision? It might
perhaps be regarded as valueless, if it did not effect an economy of logical primitives and their expressions in language.
On the assumption of two ways of judging we need:

1. assertoric force for affirmatives;
2. assertoric force for negatives, e.g. inseparably
attached to the word 'false';
3. a negative word like 'not' in sentences uttered non-
assertorically.

If on the other hand we assume only a single way of judging,
we only need:

1. assertoric force;
2. a negative word.

Such economy always shows that analysis has been pushed
further, which leads to a clearer insight. There hangs together with this an economy as regards a principle of inference; with our decision we can make do with one where

otherwise we need two. If we *can* make do with one way of judging, then we *must*; and in that case we cannot assign to one way of judging the function of setting up order and connexion, and to another, the function of dissolving this.

Thus for every thought there is a contradictory[11] thought; we acknowledge the falsity of a thought by admitting the truth of its contradictory. The sentence that expresses the contradictory thought is formed from the expression of the original thought by means of a negative word.

The negative word or syllable often seems to be more closely united to part of the sentence, e.g. the predicate. This may lead us to think that what is negated is the content, not of the whole sentence, but just of this part. We may call a man uncelebrated and thereby indicate the falsity of the thought that he is celebrated. This may be regarded as the negative answer to the question 'is the man celebrated?'; and hence we may see that we are not here just negating the sense of a word. It is incorrect to say: 'Because the negative syllable is combined with part of the sentence, the sense of the whole sentence is not negated.' On the contrary: it is by combining the negative syllable with a part of the sentence that we do negate the content of the whole sentence. That is to say: in this way we get a sentence in which there is a thought contradicting the one in the original sentence.

I do not intend by this to dispute that negation is sometimes restricted just to a part of the whole thought.

If one thought contradicts another, then from a sentence whose sense is the one it is easy to construct a sentence expressing the other. Consequently the thought that contradicts another thought appears as made up of that thought and negation. (I do not mean by this, the act of denial.) But the words 'made up of,' 'consist of,' 'component,' 'part' may lead to our looking at it the wrong way. If we choose to speak of parts in this connexion, all the same these parts

[11] We could also say 'an opposite thought.'

are not mutually independent in the way that we are elsewhere used to find when we have parts of a whole. The thought does not, by its make-up, stand in any need of completion; it is self-sufficient. Negation on the other hand needs to be completed by a thought. The two components, if we choose to employ this expression, are quite different in kind and contribute quite differently towards the formation of the whole. One completes, the other is completed. And it is by this completion that the whole is kept together. To bring out in language the need for completion, we may write 'the negation of . . .', where the blank after 'of' indicates where the completing expression is to be inserted. For the relation of completing, in the realm of thoughts and their parts, has something similar corresponding to it in the realm of sentences and their parts. (The preposition 'of', $<$'*von*'$>$, followed by a substantive can also be replaced $<$in German$>$ by the genitive of the substantive; this may as a rule be more idiomatic, but does not lend itself so well to the purpose of expressing the part that needs completion.) An example may make it even clearer what I have here in mind. The thought that contradicts the thought:

$$(21/20)^{100} \text{ is equal to } {}^{10}\sqrt{10^{21}}$$

is the thought:

$$(21/20)^{100} \text{ is not equal to } {}^{10}\sqrt{10^{21}}.$$

We may also put this as follows:

'The thought:

$$(21/20)^{100} \text{ is not equal to } {}^{10}\sqrt{10^{21}}$$

is the negation of the thought:

$$(21/20)^{100} \text{ is equal to } {}^{10}\sqrt{10^{21}}.'$$

In the last expression (after the penultimate 'is') we can see how the thought is made up of a part that needs completion and a part that completes it. From now on I shall use the word 'negation' (except, e.g., within quotation marks) always with the definite article. The definite article '*the*' in the expression

'*the* negation of the thought that 3 is greater than 5'

shows that this expression is meant to designate a definite single thing. This single thing is in our case a thought. The definite article makes the whole expression into a singular name, a proxy for a proper name.

The negation of a thought is itself a thought, and can again be used to complete *the negation*.[B] If I use, in order to complete *the negation*,[B] the negation of the thought that $(21/20)^{100}$ is equal to $^{10}\sqrt{10^{21}}$, what I get is:

the negation of the negation of the thought that $(21/20)^{100}$ is equal to $^{10}\sqrt{10^{21}}$.

This is again a thought. Designations of thoughts with such a structure are got according to the pattern:

'the negation of the negation of A,'

where 'A' takes the place of the designation of a thought. Such a designation is to be regarded as directly composed of the parts:

'the negation of ——'
and 'the negation of A.'

But it may also be regarded as made up of the parts:

'the negation of the negation of ——'
and: 'A.'

[B] I.e. to complete the thought-component whose verbal expression is '*the negation* (*of*) . . .', so as to get a complete thought; just as, in the realm of language, we get a complete designation of a thought by inserting a designation of a thought in the blank of 'the negation of——.' (The italics in the text are mine, not Frege's.)

Here I have first combined the middle part with the part that stands to the left of it and then combined the result with the part 'A' that stands to the right of it; whereas originally the middle part was combined with 'A,' and the designation so got, viz.

'the negation of A,'

was combined with what stood to the left of it

'the negation of ———.'

The two different ways of regarding the designation have answering to them two ways of regarding the structure of the thought designated.[c]

If we compare the designations:

'the negation of the negation of: $(21/20)^{100}$ is equal to $^{10}\sqrt{10^{21}}$' and the negation of the negation of: 4 is greater than 3'

we recognize a common constituent:

'the negation of the negation of ———':

this designates a part common to the two thoughts—a thought-component that stands in need of completion. In each of our two cases, it is completed by means of a thought: in the first case, the thought that $(21/20)^{100}$ is equal to $^{10}\sqrt{10^{21}}$; in the second case, the thought that 5 is greater than 3. The result of this completion is in either case a thought. This common component, which stands in need of completion, may be called double negation. This example shows how something that needs completion can be amalgamated with something that needs completion to form something that needs completion. Here we are presented with a singular case; we have something—the negation of . . .—amalgamated

---

[c] *Bezeichnenden* is here surely a misprint for *bezeichneten* or *zu bezeichnenden*.

with itself. Here, of course, metaphors derived from the corporeal realm fail us; for a body cannot be amalgamated with itself so that the result is something different from it. But then neither do bodies need completion, in the sense I intend here. Congruent bodies *can* be put together; and in the realm of designations we have congruence in our present case. Now what corresponds to congruent designations is one and the same thing in the realm of designata.

Metaphorical expressions, if used cautiously, may after all help towards an elucidation. I compare that which needs completion to a wrapping, e.g. a coat, which cannot stand upright by itself; in order to do that, it must be wrapped round somebody. The man whom it is wrapped round may put on another wrapping, e.g. a cloak. The two wrappings unite to form a single wrapping. There are thus two possible ways of looking at the matter; we may say either that a man who already wore a coat was now dressed up in a second wrapping, a cloak, or, that his clothing consists of two wrap-pings—coat and cloak. These ways of looking at it have absolutely equal justification. The additional wrapping always combines with the one already there to form a new wrapping. Of course we must never forget in this connexion that dressing up and putting things together are processes in time, whereas what corresponds to this in the realm of thoughts is timeless.

If A is a thought not belonging to fiction, the negation of A likewise does not belong to fiction. In that case, of the two thoughts: A and the negation of A: there is always one and only one that is true. Likewise, of the two thoughts: the negation of A, and the negation of the negation of A: there is always one and only one that is true. Now the negation of A is either true or not true. In the first case, neither A nor the negation of the negation of A is true. In the second case, both A and the negation of the negation of A are true. Thus of the two thoughts: A, and the negation of the negation of

A: either both are true or neither is. I may express this as follows:

Wrapping up a thought in double negation does not alter its truth-value.

Compound Thoughts

It is astonishing what language can do. With a few syllables it can express an incalculable number of thoughts, so that even a thought grasped by a terrestrial being for the very first time can be put into a form of words which will be understood by someone to whom the thought is entirely new. This would be impossible, were we not able to distinguish parts in the thought corresponding to the parts of a sentence, so that the structure of the sentence serves as an image of the structure of the thought. To be sure, we really talk figuratively when we transfer the relation of whole and part to thoughts; yet the analogy is so ready to hand and so generally appropriate that we are hardly ever bothered by the hitches which occur from time to time.

If, then, we look upon thoughts as composed of simple parts, and take these, in turn, to correspond to the simple parts of sentences, we can understand how a few parts of sentences can go to make up a great multitude of sentences, to which, in turn, there correspond a great multitude of thoughts. But the question now arises how a thought comes to be constructed, and how its parts are so combined together that the whole amounts to something more than the parts taken separately. In my article 'Negation' I considered the case where a thought seems to be composed of a part needing completion (the unsaturated part, as one may also call it, represented in language by the negating word), and a thought. There can be no negation without something negated, and this is a thought. The whole owes its unity to

E

the fact that the thought saturates the unsaturated part or, as we can also say, completes the part needing completion. And it is natural to suppose that, for logic in general, combination into a whole always comes about by saturation of something unsaturated.[12]

But here a special case of such combination is to be considered, namely that in which two thoughts are combined to form a single thought. In the realm of language, this is represented by the combination of two sentences into a whole that likewise is a sentence. On the analogy of the grammatical term 'compound sentence', I shall employ the expression 'compound thought', without wishing to imply by this that every compound sentence has a compound thought as its sense, or that every compound thought is the sense of a compound sentence. By 'compound thought' I shall understand a thought consisting of thoughts, but not of thoughts alone. For a thought is complete and saturated, and needs no completion in order to exist. For this reason, thoughts do no cleave to one another unless they are connected together by something that is not a thought, and it may be taken that this connective is unsaturated. The compound thought must itself be a thought: that is, something either true or false (*tertium non detur*).

Not every sentence composed, linguistically speaking, of sentences will provide us with a serviceable example; for grammar recognizes sentences which logic cannot acknowledge as sentences proper because they do not express thoughts. This is illustrated in relative clauses; for in a relative clause detached from its main clause, we cannot tell what the relative pronoun is supposed to refer to. Such a clause contains no sense whose truth can be investigated; in other words, the sense of a detached relative clause is not a thought. So we must not expect a compound sentence con-

---

[12] Here, and in what follows, it must always be remembered that this saturation and combination are not temporal processes.

sisting of a main and a relative clause to have as its sense a compound thought.

FIRST KIND OF COMPOUND THOUGHT

In language, the simplest case seems to be that of two main clauses conjoined by 'and'. But the matter is not so simple as it first appears, for in an assertoric sentence we must distinguish between the thought expressed and assertion. Only the former is in question here, for it is not acts of judgement that are to be conjoined.[13] I therefore take the sentences conjoined by 'and' to be uttered without assertive force. Assertive force can most easily be eliminated by changing the whole into a question; for one can express the same thought in a question as in an assertoric sentence, only without asserting it. If we use 'and' to conjoin two sentences, neither of which is uttered with assertive force, then we have to ask whether the sense of the resultant whole is a thought. For not only each of the component sentences, but also the whole, must have a sense which can be made the content of a question. Suppose witnesses are asked: 'Did the accused deliberately set fire to the pile of wood, and deliberately start a forest-fire?'; the problem then arises whether two questions are involved here, or only one. If the witnesses are free to reply affirmatively to the question about the pile of wood, but negatively to that about the forest-fire, then we have two questions, each containing a thought, and there is no question of a single thought compounded out of these two. But if—as I shall suppose—the witnesses are permitted to answer only 'yes' or 'no', without dividing the whole into

[13] Logicians often seem to mean by 'judgement' what I call 'thought'. In my terminology, one judges by acknowledging a thought as true. This act of acknowledgement I call 'judgement'. Judgement is made manifest by a sentence uttered with assertive force. But one can grasp and express a thought without acknowledging it as true, i.e. without judging.

sub-questions, then this whole is a single question which should be answered affirmatively only if the accused acted deliberately both in setting fire to the pile of wood and also in starting the forest-fire; and negatively in any other case. Thus, a witness who thinks that the accused certainly set fire to the pile of wood on purpose, but that the fire then spread further and set the forest alight without his meaning it to, must answer the question in the negative. For the thought in the whole question must be distinguished from the two component thoughts: it contains, as well as the component thoughts, that which combines them together; and this is represented in language by the word 'and'. This word is used here in a particular way; we are concerned only with its use as a conjunction between two sentences proper. I call any sentence a sentence proper if it expresses a thought. But a thought is something which must be either true or false, *tertium non datur*. Furthermore, the 'and' now under discussion can only conjoin sentences which are uttered non-assertively. I do not mean by this to exclude the act of judgement; but if it occurs, it must relate to the compound thought as a whole. If we wish to present a compound of this first kind as true, we may use the phrase 'It is true that ... and that...'.

Our 'and' is not meant to conjoin interrogative sentences, any more than assertoric sentences. In our example the witnesses are confronted with only one question. But the thought proposed for judgement by this question is composed of two thoughts. In his reply, however, the witness must give only a single judgement. Now this may certainly seem an artificial refinement; for doesn't it really come to the same thing, whether the witness first replies affirmatively to the question 'Did the accused deliberately set fire to the pile of wood?' and then to the question 'Did the accused deliberately start a forest-fire?', or rather replies affirmatively at one stroke to the whole question? This may well seem so, in case of an affirmative reply, but the difference shows up more clearly where the answer is negative. For this reason it is useful to express the thought in a single question, since then both negative and affirmative cases will have to be considered in order to understand the thought correctly.

The 'and' whose mode of employment is more precisely delimited in this way seems doubly unsaturated: to saturate it we require both a sentence preceding and another following. And what corresponds to 'and' in the realm of sense must also be doubly unsaturated: inasmuch as it is saturated by thoughts, it combines them together. As a mere thing, of course, the group of letters 'and' is no more unsaturated than any other thing.[14] It may be called unsaturated in respect of its employment as a symbol meant to express a sense, for here it can have the intended sense only when situated between two sentences: its purpose as a symbol requires completion by a preceding and a succeeding sentence. It is really in the realm of sense that unsaturatedness is found, and it is transferred from there to the symbol.

If 'A' and 'B' are both sentences proper, uttered with neither assertive nor interrogative force, then 'A and B' is likewise a sentence proper, and its sense is a compound thought of the first kind. Hence I also say that 'A and B' expresses a compound thought of the first kind.

That 'B and A' has the same sense as 'A and B' we may see without proof by merely being aware of the sense. Here we have a case where two linguistically different expressions correspond to the same sense. This divergence of expressive symbol and expressed thought is an inevitable consequence of the difference between spatio-temporal phenomena and the world of thoughts.[15]

Finally, we may point out an inference that holds in this connexion:

14 Cf. p. 56.
15 Another case of this sort is that 'A and A' has the same sense as 'A'.

A is true;[16]
B is true; therefore
(A and B) is true.

SECOND KIND OF COMPOUND THOUGHT

The negation of a compound of the first kind between one thought and another is itself a compound of the same two thoughts. I shall call it a compound thought of the second kind. Whenever a compound thought of the first kind out of two thoughts is false, the compound of the second kind out of them is true, and conversely. A compound of the second kind is false only if each compounded thought is true, and a compound of the second kind is true whenever at least one of the compounded thoughts is false. In all this it is assumed throughout that the thoughts do not belong to fiction. By presenting a compound thought of the second kind as true, I declare the compounded thoughts to be non-conjoinable.

Without knowing whether

$$(21/20)^{100} \text{ is greater than } \sqrt[10]{10^{21}},$$

or whether

$$(21/20)^{100} \text{ is less than } \sqrt[10]{10^{21}},$$

I can still recognize that the compound of the first kind out of these two thoughts is false. Accordingly, the corresponding compound of the second kind is true. Apart from the thoughts compounded, we have something that connects them, and here too the connective is doubly unsaturated: the connexion comes about in that the component thoughts saturate the connective.

[16] When I write 'A is true', I mean more exactly 'the thought expressed in the sentence "A" is true'. So too in analogous cases.

To express briefly a compound thought of this kind, I write

'not [A and B]',

where 'A' and 'B' are the sentences corresponding to the compounded thoughts. The connective stands out more clearly in this expression: it is the sense of whatever occurs in the expression apart from the letters 'A' and 'B'. The two gaps in the expression

'not [and   ]'

bring out the two-fold unsaturatedness. The connective is the doubly unsaturated sense of this doubly unsaturated expression. By filling the gaps with expressions of thoughts, we form the expression of a compound thought of the second kind. But we really should not talk of the compound thought as originating in this way, for it is a thought and a thought does not originate.

In a compound thought of the first kind, the two thoughts may be interchanged. The same interchangeability must also hold for the negation of a compound thought of the first kind, hence for a compound thought of the second kind. If, therefore, 'not [A and B]' expresses a compound thought, then 'not [B and A]' expresses the same compound of the same thoughts. This interchangeability should no more be regarded as a theorem here than for compounds of the first kind, for there is no difference in sense between these expressions. It is therefore self-evident that the sense of the second compound sentence is true if that of the first is true— for it is the same sense.

An inference may also be mentioned for the present case:

not [A and B] is true;
A is true; therefore
B is false.

THIRD KIND OF COMPOUND THOUGHT

A compound of the first kind, formed from the negation of one thought conjoined with the negation of another thought, is also a compound of these thoughts themselves. I call it a compound of the third kind out of the first thought and the second. Let the first thought, for example, be that Paul can read, and the second that Paul can write; then the compound of the third kind out of these two thoughts is the thought that Paul can neither read nor write. A compound thought of the third kind is true only if each of the two compounded thoughts is false, and it is false if at least one of the compounded thoughts is true. In compound thoughts of the third kind, the component thoughts are also interchangeable. If 'A' expresses a thought, then 'not A' must express the negation of this thought, and similarly for 'B'. Hence, if 'A' and 'B' are genuine sentences, then the sense of

'(not A) and (not B)',

for which I also write

'neither A nor B'

is the compound of the third kind out of the two thoughts expressed by 'A' and 'B'.

Here the connective is the sense of everything in these expressions apart from the letters 'A' and 'B'. The two gaps in

'(not) and (not)',

or in

'neither , nor ',

indicate the two-fold unsaturatedness of these expressions

which corresponds to the two-fold unsaturatedness of the connective. When the latter is saturated by thoughts, there comes about the compound of the third kind out of these thoughts.

Once again we may mention an inference:

> A is false;
> B is false; therefore
> (neither A nor B) is true.

The brackets are to make it clear that what they contain is the whole whose sense is presented as true.

FOURTH KIND OF COMPOUND THOUGHT

The negation of a compound of the third kind between two thoughts is likewise a compound of these two thoughts: it may be called a compound thought of the fourth kind. A compound of the fourth kind out of two thoughts is a compound of the second kind out of the negations of these thoughts. In presenting such a compound thought as true, we thereby assert that at least one of the compounded thoughts is true. A compound thought of the fourth kind is false only if each of the compounded thoughts is false. Given once again that 'A' and 'B' are sentences proper, the sense of

'not [(not A) and (not B)]'

is a compound thought of the fourth kind between the thoughts expressed by 'A' and 'B'. The same holds of

'not [neither A nor B]',

which may be written more briefly

'A or B'.

Taken in this sense, 'or' occurs only between sentences—

indeed only between sentences proper. By recognizing the truth of such a compound thought, I do not rule out the truth of both compounded thoughts: we have in this case the non-exclusive 'or'. The connective is the sense of whatever occurs in 'A' or 'B' apart from 'A' and 'B', that is, the sense of

'(or)'

where the gaps on both sides of 'or' indicates the two-fold want of fulfilment in the connective. The sentences conjoined by 'or' should be regarded merely as expressions of thoughts, and not therefore as individually endowed with assertive force. The compound thought as a whole, on the other hand, may be acknowledged as true. The linguistic expression does not make this clear: each component sentence of the assertion '5 is less than 4, or 5 is greater than 4' has the linguistic form which it would also have if it were uttered separately with assertive force, whereas really only the whole compound is meant to be presented as true.

Perhaps it will be found that the sense here assigned to the word 'or' does not always agree with ordinary usage. On this point it should first be noted that in determining the sense of scientific expressions we cannot undertake to concur exactly with the usage of ordinary life; this, indeed, is for the most part unsuited to scientific purposes, where we feel the need for more precise definition. The scientist must be allowed to diverge, in his use of the word 'ear', from what is otherwise the custom. In the field of logic, reverberations of side-thoughts may be distracting. In virtue of what has been said about our use of 'or', it can truly be asserted 'Frederick the Great won the battle of Rossbach, or two is greater than three'. This leads someone to think: 'Good Heavens! What does the battle of Rossbach have to do with the nonsense that two is greater than three?' But 'two is greater than three' is false, not nonsense: it makes no difference to logic whether the falsity of a thought is easy or diffi-cult to discern. In sentences conjoined by 'or', we usually suppose that the sense of the one has something to do with that of the other, that there is some sort of relationship be-tween them. Such a relationship may well indeed be specifi-able for a given case, but for different cases there will be different relationships, and it will therefore be impossible to specify a relationship of meaning which would always be attached to 'or' and could accordingly count as going with the sense of this word. 'But why does the speaker add the second sentence at all? If he wants to assert that Frederick the Great won the battle of Rossbach, then surely the first sentence would be sufficient. We may certainly assume that he does not want to claim that two is greater than three; and if he had been satisfied with just the first sentence, he would have said more with fewer words. Why, therefore, this waste of words?' These questions, too, only distract us into side-issues. Whatever may be the speaker's intentions and motives for saying just this and not that, our concern is not with these at all, but solely with what he says.

Compound thoughts of the first four kinds have this in common, that their component thoughts may be inter-changed.

Here, too, follows another inference:

(A or B) is true;
A is false; therefore
B is true.

FIFTH KIND OF COMPOUND THOUGHT

By forming a compound of the first kind out of the nega-tion of one thought and a second thought, we get a com-pound of the fifth kind out of these two thoughts. Given

that 'A' expresses the first thought and 'B' expresses the
second, the sense of

$$\text{`(not A) and B'}$$

is such a compound thought. A compound of this kind is
true if, and only if, the first compounded thought is false
while the second is true. Thus, for example, the compound
thought expressed by

$$\text{`(not } 3^2 = 2^3) \text{ and } (2^4 = 4^3)\text{'}$$

viz. the thought that 3^2 is not equal to 2^3 and 2^4 is equal to
4^2, is true. After seeing that 2^4 is equal to 4^2, someone may
think that in general the exponent of a number raised to a
power can be interchanged with the number itself. Someone
else may then try to correct this mistake by saying '2^4 equals
4^2 but 2^3 does not equal 3^2'. If it is asked what difference
there is between conjunction with 'and' and with 'but', the
answer is: with respect to what I have called the 'thought'
or the 'sense' of the sentence, it is immaterial whether the
idiom of 'and' or that of 'but' is chosen. The difference
comes out only in what I call the light cast on the thought
(cf. my article—'Thoughts') and does not belong to the
province of logic.

The connective in a compound thought of the fifth kind
is the doubly incomplete sense of the doubly incomplete
expression

$$\text{`(not) and ( )'}.$$

Here the compounded thoughts are not interchangeable, for

$$\text{`(not B) and A'}$$

does not express the same as

$$\text{`(not A) and B'}.$$

The first thought does not occupy the same kind of position

tate to coin a new word, I am obliged to use the word 'position' with a transferred meaning. In speaking of written expressions of thoughts, 'position' may be taken to have its ordinary spatial meaning. But a position in the expression of a thought must correspond to something in the thought itself, and for this I shall retain the word 'position'. In the present case we cannot simply make the two thoughts exchange their position, but we can set the negation of the second thought in the position of the first, and at the same time the negation of the first in the position of the second. (Of course, this too must be taken with a grain of salt, for an operation in space and time is not intended.) Thus from

'(not A) and B'

we obtain

'(not (not B)) and (not A)'.

But since 'not (not B)' has the same sense as 'B', we have here

'B and (not A)',

which expresses the same as

'(not A) and B'.

Sixth Kind of Compound Thought

By negating a compound of the fifth kind out of two thoughts, we get a compound of the sixth kind out of the same two thoughts. We can also say that a compound of the second kind out of the negation of one thought and a second thought is a compound of the sixth kind out of these two thoughts. A compound of the fifth kind is true if and only if its first component thought is false, but the second is true. From this it follows that a compound of the sixth kind out of two thoughts is false if and only if its first component is false, but the second is true. Such a compound thought is therefore true given only the truth of its first component thought, regardless of whether the second is true or false. It is also true given only the falsehood of its second component thought, regardless of whether the first is true or false.

Without knowing whether

$$((21/20)^{100})^2 \text{ is greater than } 2^2,$$

or whether

$$(21/20)^{100} \text{ is greater than } 2,$$

I can still recognize as true the compound of the sixth kind out of these two thoughts. The negation of the first thought excludes the second thought, and *vice versa*. We can put it as follows:

'If $(21/20)^{100}$ is greater than 2,
then $((21/20)^{100})^2$ is greater than 2^2.'

Instead of 'compound thought of the sixth kind', I shall also speak of 'hypothetical compound thought', and I shall refer to the first and second components of a hypothetical compound thought as 'consequent' and 'antecedent' respectively. Thus, a hypothetical compound thought is true if its consequent is true; it is also true if its ancedent is false, regardless of whether the consequent is true or false. The consequent must always be a thought.

Given once again that 'A' and 'B' are sentences proper, then

'not (not A) and B'

expresses a hypothetical compound with the sense (thought-content) of 'A' as consequent and the sense of 'B' as antecedent. We may also write instead:

'If B, then A'.

But here, indeed, doubts may arise. It may perhaps be maintained that this does not square with linguistic usage. In reply, it must once again be emphasized that science has to be allowed its own terminology, that it cannot always bow to ordinary language. Just here I see the greatest difficulty for philosophy: the instrument it finds available for its work, namely ordinary language, is little suited to the purpose, for its formation was governed by requirements wholly different from those of philosophy. So also logic is first of all obliged to fashion a usable instrument from those already to hand. And for this purpose it initially finds but little in the way of usable instruments available.

Many would undoubtedly declare that the sentence

'If 2 is greater than 3, then 4 is a prime number'

is nonsense; and yet, according to my stipulation it is true because the antecedent is false. To be false is not yet to be nonsense. Without knowing whether

$$\sqrt[10]{10^{21}} \text{ is greater than } (21/20)^{100},$$

we can see that

$$\text{If } \sqrt[10]{10^{21}} \text{ is greater than } (21/20)^{100},$$
$$\text{then } (\sqrt[10]{10^{21}})^2 \text{ is greater than } ((21/20)^{100})^2;$$

and nobody will see any nonsense in that. But it is false that

$$\sqrt[10]{10^{21}} \text{ is greater than } (21/20)^{100},$$

and it is equally false that

$$(\sqrt[10]{10^{21}})^2 \text{ is greater than } ((21/20)^{100})^2.$$

If this could be seen as easily as the falsity of '2 is greater than 3', then the hypothetical compound thought of the

present example would seem just as nonsensical as that of
the previous one. Whether the falsity of a thought can be
seen with greater or less difficulty is of no matter from a
logical point of view, for the difference is a psychological
one.

The thought expressed by the compound sentence

'If I own a cock which has laid eggs today, then
Cologne Cathedral will collapse tomorrow morning'

is also true. Someone will perhaps say: 'But here the antecedent has no inner connexion at all with the consequent'.
In my account, however, I required no such connexion, and
I ask that 'If B, then A' should be understood solely in
terms of what I have said and expressed in the form

'not [(not A) and B]'.

It must be admitted that this conception of a hypothetical
compound thought will at first be thought strange. But my
account is not designed to square with ordinary linguistic
usage, which is generally too vague and ambiguous for the
purposes of logic. Questions of all kinds arise at this point,
e.g. the relation of cause and effect, the intention of a speaker
who utters a sentence of the form 'If B, then A', the grounds
on which he holds its content to be true. The speaker may
perhaps give hints in regard to such questions arising among
his hearers. These hints are among the adjuncts which often
surround the thought in ordinary language. My task here is
to remove the adjuncts and thereby to pick out, as the logical kernel, a compound of two thoughts, which I have
called a hypothetical compound thought. Insight into the
structure of thoughts compounded of two thoughts must
provide the foundation for consideration of multiply compound thoughts.

What I have said about the expression 'If B, then A' must
not be so understood as to imply that every compound sen-

tence of this form expresses a hypothetical compound thought. If either 'A' or 'B' by itself does not completely express a thought, and is not therefore a genuine sentence, the case is altered. In the compound sentence,

'If someone is a murderer, then he is a criminal',

neither the antecedent-clause nor the consequent-clause, taken by itself, expresses a thought. Without some further clue, we cannot determine whether what is expressed in the sentence 'He is a criminal' is true or false when detached from this compound; for the word 'he' is not a proper name, and in the detached sentence it designates nothing. It follows that the consquent-clause expresses no thought, and is therefore not a genuine sentence. This holds of the antecedent-clause as well, for it likewise has a non-designating component, namely 'someone'. Yet the compound sentence can none the less express a thought. The 'someone' and the 'he' refer to each other. Hence, and in virtue of the 'If—, then—', the two clauses are so connected with one another that they together express a thought; whereas we can distinguish three thoughts in a hypothetical compound thought, namely the antecedent, the consequent, and the thought compounded of these. Thus, compound sentences do not always express compound thoughts, and it is very important to distinguish the two cases which arise for compound sentences of the form

'If B, then A'.

Once again I append an inference:

[If B, then A] is true;
B is true; therefore
A is true.

In this inference, the characteristic feature of hypothetical

F

compound thoughts stands out, perhaps in its clearest form. The following mode of inference is also noteworthy:

[If C, then B] is true;
[If B, then A] is true; therefore
[If C, then A] is true.

I should like here to call attention to a misleading way of speaking. Many mathematical writers express themselves as if conclusions could be drawn from a thought whose truth is still doubtful. In saying 'I infer A from B', or 'I conclude from B that A is true', we take B for one of the premises or the sole premise of the inference. But before recognizing its truth, one cannot use a thought as premise of an inference, nor can one infer or conclude anything from it. If anyone still thinks this can be done, he is apparently confusing recognition of the truth of a hypothetical compound thought with performing an inference in which the antecedent of this compound is taken for a premise. Now recognition of the truth of the sense of

'If C, then A'

can certainly depend on an inference, as in the example given above, while there may yet be a doubt about the truth of C.[17] But in this case, the thought expressed by 'C' is by no means a premise of the inference; the premise, rather, was the sense of the sentence

'If C, then B'.

If the thought-content of 'C' were a premise of the inference, then it would not occur in the conclusion: for that is just how inference works.

We have seen how, in a compound thought of the fifth kind, the first thought can be replaced by the negation of

---

[17] More precisely: whether the thought expressed by 'C' is true.

the second, and the second simultaneously by the negation of the first, without altering the sense of the whole. Now since a compound thought of the sixth kind is the negation of a compound thought of the fifth kind, the same also holds for it: that is, we can replace the antecedent of a hypothetical compound by the negation of the consequent, and the consequent simultaneously by the negation of the antecedent, without thereby altering its sense. (This is contraposition, the transition from *modus ponens* to *modus tollens*.)

SUMMARY OF THE SIX COMPOUND THOUGHTS

I.	A and B;	II. not (A and B);
III.	(not A) and (not B);	IV. not ((not A) and (not B));
V.	(not A) and B;	VI. not ((not A) and B).

It is tempting to add

A and (not B).

But the sense of 'A and (not B)' is the same as that of '(not B) and A', for any genuine sentences 'A' and 'B'. And since '(not B) and A' has the same form as '(not A) and B', we get nothing new here, but only another expression of a compound thought of the fifth kind; and in 'not (A and (not B))' we have another expression of a compound thought of the sixth kind. Thus our six kinds of compound thought form a completed whole, whose primitive elements seem here to be the first kind of compound and negation. However acceptable to psychologists, this apparent pre-eminence of the first kind of compound over the others has no logical justification; for any one of the six kinds of compound thought can be taken as fundamental and can be used, together with negation, for deriving the others; so that, for logic, all six kinds have equal justification. If, for example, we start with the hypothetical compound

If B, then C, i.e. not ((not C) and B)),

and replace 'C' by 'not A', then we get

If B, then not A, i.e. not (A and B).

By negating the whole, we get

not (if B, then not A), i.e. A and B.

from which it follows that

'not (if B, then not A)'

says the same as

'A and B'.

We have thereby derived a compound of the first kind from a hypothetical compound and negation; and since compounds of the first kind and negation together suffice for the derivation of the other compound thoughts, it follows that all six kinds of compound thought can be derived from hypothetical compounds and negation. What has been said of the first and the sixth kinds of compound holds in general of all our six kinds of compound thought, so that none has any priority over the others. Each of them can serve as a basis for deriving the others, and our choice is not governed by any fact of logic.

A similar situation exists in the foundation of geometry. Two different geometries can be formulated in such a way that certain theorems of the one occur as axioms of the other, and conversely.

Let us now consider cases where a thought is compounded with itself rather than with some different thought. For any genuine sentence 'A', 'A and A' expresses the same thought as 'A': the former says no more and no less than the latter. It follows that 'not (A and A)' expresses the same as 'not A'.

Equally, '(not A) and (not A)' also expresses the same as

'not A'; and consequently 'not [(not A) and (not A)]' also expresses the same as 'not (not A), or 'A'. Now, 'not [(not A) and (not A)]' epresss a compound of the fourth kind and instead of this we can say 'A or A'. Accordingly, not only 'A and A', but also 'A or A' has the same sense as 'A'.

It is otherwise for compounds of the fifth kind. The compound thought expressed by '(not A) and A' is false, since, of two thoughts, where one is the negation of the other, one must always be false; so that a compound of the first kind composed of them is likewise false. The compound of the sixth kind out of a thought and itself, namely that expressed by 'not [(not A) and A]' is accordingly true (assuming that 'A' is a genuine sentence). We can also render this compound thought verbally by the expression 'If A, then A'; for example, 'If the Schneekoppe is higher than the Brocken, then the Schneekoppe is higher than the Brocken'.

In such a case the questions arise: 'Does this sentence express a thought? Doesn't it lack content? Do we learn anything new upon hearing it?' Now it may happen that before hearing it someone did not know this truth at all, and had therefore not acknowledged it. To that extent one could, under certain conditions, learn something new from it. It is surely an undeniable fact that the Schneekoppe is higher than the Brocken if the Schneekoppe is higher than the Brocken. Since only thoughts can be true, this compound sentence must express a thought; and, despite its apparent senselessness, the negation of this thought is also a thought. It must always be borne in mind that a thought can be expressed without being asserted. Here we are concerned just with thoughts, and the appearance of senselessness arises only from the assertive force with which one involuntarily takes the sentence to be uttered. But who says that anyone uttering it non-assertively does so in order to present its content as true? Perhaps he is doing it with precisely the opposite intention.

This can be generalized. Let 'O' be a sentence which expresses a particular instance of a logical law, but which is not presented as true. Then it is easy for 'not O' to seem senseless, but only because it is thought of as uttered assertively. The assertion of a thought which contradicts a logical law can indeed appear, if not senseless, then at least absurd; for the truth of a logical law is immediately evident of itself, from the sense of its expression. But a thought which contradicts a logical law may be expressed, since it may be negated. 'O' itself, however, seems almost to lack content.

Any compound thought, being itself a thought, can be compounded with other thoughts. Thus, the compound expressed by '(A and B) and C' is composed of the thoughts expressed by 'A and B' and 'C'. But we can also treat it as composed of the thoughts expressed by 'A' and 'B' and 'C'. In this way compound thoughts containing three thoughts can originate.[18] Other examples of such compounds are expressed by:

'not [(not A) and (B and C)]', and
'not [(not A) and ((not B) and (not C))]'.

So too it will be possible to find examples of compound thoughts containing four, five, or more thoughts.

Compound thoughts of the first kind, and negation, are together adequate for the formation of all these compounds, and any other of our six kinds of compound can be chosen instead of the first. Now the question arises whether every compound thought is formed in this way. So far as mathematics is concerned, I am convinced that it includes no compound thoughts formed in any other way. It will scarcely be otherwise in physics, chemistry, and astronomy as well; but 'in order that' clauses call for caution and seem to require more precise investigation. Here I shall leave this question

---

[18] This origination must not be regarded as a temporal process.

open. Compound thoughts thus formed with the aid of negation from compounds of the first kind seem, at all events, to merit a special title. They may be called mathematical compound thoughts. This should not be taken to mean that there are compound thoughts of any other type. Mathematical compound thoughts seem to have something else in common; for if a true component of such a compound is replaced by another true thought, the resultant compound thought is true or false according to whether the original compound is true or false. The same holds if a false component of a mathematical compound thought is replaced by another false thought. I now want to say that two thoughts have the same truth-value if they are either both true or both false. I maintain, therefore, that the thought expressed by 'A' has the same truth-value as that expressed by 'B' if either 'A and B' or else '(not A) and (not B)' expresses a true thought. Having established this, I can phrase my thesis in this way:

'If one component of a mathematical compound thought is replaced by another thought having the same truth-value, then the resultant compound thought has the same truth-value as the original.'

Select Bibliography

This bibliography is based on the one compiled by Professor G. Patzig for his own German edition of Frege's *Logische Untersuchungen* (Gottingen: Vandenhoek & Rupprecht, 1976). I am grateful for his kind permission to use his work.

A General Bibliography. *Appended to*
1 Dummett, M.: *Frege: Philosophy of Language.* London 1973. 685–93.

B Introductory material
2 Passmore, John: *A Hundred Years of Philosophy.* London 1957, 1967, pp. 149–57.
3 Kneale, William C.: 'Gottlob Frege and Mathematical Logic', in *Revolution in Philosophy*, ed. G. Ryle. London 1960, pp. 26–40.
4 Geach, Peter Thomas: 'Gottlob Frege', in: G. E. M. Anscombe & P. T. Geach, *Three Philosophers: Aristotle, Aquinas, Frege.* Oxford 1961, pp. 127–62.
5 Kneale, William C.: 'Frege's General Logic', in: W. Kneale and M. Kneale, *The Development of Logic.* Oxford 1962, pp. 478–512.
6 Thiel, Christian: *Sinn und Bedeutung in der Logik Gottlob Freges.* Monographien zur philosophischen Forschung, Vol. 43. Meisenheim/Glan 1965, pp. VIII, 172.
7 Dummett, Michael: 'Gottlob Frege', in: *The Encyclopedia of Philosophy*, ed. Paul Edwards et al. New York, London 1967, Vol. III, pp. 225–37.
8 Klemke, E. D. (ed.): *Essays on Frege.* Urbana, Chicago & London 1968, pp. XIV, 586.

9 Largeault, Jean: *Logique et philosophie chez Frege.* Paris-Louvain 1970, pp. XXVII, 486.

10 Patzig, Gunther: 'Gottlob Frege und die logische Analyse der Sprache', in: G. Patzig: *Sprache und Logik.* Gottingen 1970, pp. 77–100.

11 Dummett, Michael: See (1) above.

C   Special works connected with the topic of this volume:

12 Krenz, Editha: *Der Zahlbegriff bei Frege.* Doctoral dissertation, Vienna 1942, pp. 116ff.

13 Bierich, Marcus: *Freges Lehre von dem Sinn und der Bedeutung der Urteile und Russells Kritik an dieser Lehre.* Doctoral dissertation (typescript) Hamburg 1951.

14 Marshall, William: 'Frege's Theory of Functions and Objects'. *The Philosophical Review*, 62, 1953, pp. 374–390. Reprinted in (8), pp. 249–67.

15 Dummett, Michael: 'Frege on Functions: a Reply'. *The Philosophical Review*, 64, 1955, 96–107. Reprinted in (8), pp. 268–83.

16 Geach, Peter Thomas: 'Class and Concept'. *The Philosophical Review*, 64, 1955. Reprinted in (8), pp. 284–94.

17 Dummett, Michael: 'Note: Frege on Functions'. *The Philosophical Review*, 65, 1955, pp. 229–30. Reprinted in (8), pp. 295–7.

18 Marshall, William: 'Sense and Reference: A Reply'. *The Philosophical Review*, 65, 1956, pp. 342–61. Reprinted in (8), pp. 298–320.

19 Khatchadourian, Haig: 'Frege on Concepts'. *Theoria*, 22, 1956, pp. 85–100.

20 Searle, John R.: 'Russell's Objections to Frege's Theory of Sense and Reference'. *Analysis*, 18, 1957/58, pp. 137–43, Reprinted in (8), pp. 337–45.

21 Dummett, Michael: 'Truth'. *Proceedings of the Aristotelian Society,* 59, 1958–9, pp. 141–62.

22 Kauppi, R.: *Über Sinn, Bedeutung und Wahrheitswert der Sätze.* Acta Academiae Paedagogicae Jyväkyläensis 17, 1959, pp. 205–13.

23 Jackson, Howard: 'Frege's Ontology'. *The Philosophical Review*, 69, 1960, pp. 394–5. Reprinted in (8), pp. 77–8.

24 Caton, Charles E.: 'An Apparent Difficulty in Frege's Ontology'. *The Philosophical Review*, 71, 1962, pp. 462–475. Reprinted in: (8), pp. 99–112.

25 Fisk, Milton: 'A Paradox in Frege's Semantics'. *Philosophical Studies*, 14, 1963, pp. 56–63. Reprinted in (8), pp. 390–2.

26 Martin, R. M.: 'On the Frege–Church Theory of Meaning'. *Philosophy and Phenomenological Research*, 23, 1963, pp. 605–9.

27 Resnik, Michael D.: 'Frege's Theory of Incomplete Entities', *Philosophy of Science*, 32, 1965, pp. 329–41.

28 Parsons, Charles D.: 'Frege's Theory of Number', in: *Philosophy in America*, ed. by M. Black, London 1965, pp. 180–203.

29 Angelelli, Ignacio: 'On Identity and Interchangeability in Leibnitz and Frege'. *Notre Dame Journal of Formal Logic*, 8, 1967, pp. 94–100.

30 Linsky, Leonard: *Referring*. London, New York 1967, Chap. III: 'Sense and Reference', pp. 22–38 and Appendix, pp. 39–48.

31 Martin, R. M.: 'On Proper Names and Frege's Darstellungs-weise'. *The Monist*, 51, 1967, 1–8.

32 Schorr, Karl Eberhard: 'Der Begriff bei Kant und Frege' *Kantstudien*, 58, 1967, 227–46.

33 Gram, Moltke S.: 'Frege, Concepts and Ontology,' in (8), 1968, pp. 178–99.

34 Berka, Karel und Lothar Kreiser: 'Eine grundsätzliche Erweiterung der Semantik G. Freges'. *Deutsche Zeitschrift für Philosophie*, 16, 1968, pp. 1228–39.

35 Klemke, E. D.: 'Frege's Ontology: Realism', in (8), 1968, pp. 157–77.

36 Suter, Ronald: 'Frege and Russell on the "Paradox of Identity".' *Proceedings of the Seventh Inter-American Congress of Philosophy, 1967*, Vol. II, Quebec 1968, pp. 30–6.

37   Dudman, V. H.: 'A Note on Frege on Sense'. *The Australasian Journal of Philosophy* 47, 1969, pp. 119–22.

38   Dudman, V. H.: 'Frege's Judgement-Stroke'. *The Philosophical Quarterly*, 20, 1970, pp. 150–62.

39   Gabriel, Gottfried: 'G. Frege über semantische Eigenschaften der Dichtung. *Linguistische Berichte*, Fasc. 8, 1970, pp. 10–17.

40   Tugendhat, Ernst: 'The Meaning of "Bedeutung" in Frege'. *Analysis*, 30, 1970, pp. 177–89.

41   Welding, S. O.: 'Frege's Sense and Reference Related to Russell's Theory of Definite Descriptions'. *Revue internationale de Philosophie* 25, 1971, 389–402.

42   Dudman, V. H.: 'Frege on Assertion'. *The Philosophical Quarterly*, 22, 1972, pp. 61–4.

43   Hoche, Hans-Ulrich: 'Kritische Bemerkungen zu Freges Bedeutungslehre'. *Zeitschrift für philosophische Forschung*, 27, 1973, pp. 205–21.

44   Fabian, Reinhard: 'Sinn und Bedeutung von Namen und Sätzen. Eine Untersuchung zur Semantik Gottlob Freges'. Doctoral dissertation. Graz, 1974.

45   Stuhlmann-Laeisz, R.: 'Freges Auseinandersetzung mit der Auffassung von "Existenz" als einem Prädikat der ersten Stufe und Kants Argumentation gegen den ontologischen Gottesbeweis, in: Christian Thiel (ed.), *Frege und die moderne Grundlagenforschung*, Meisenheim/Glan, 1975, pp. 119–33.

46   Prauss, Gerold: 'Freges Beitrag zur Erkenntnistheorie. Überlegungen zu seinem Aufsatz "Der Gedanke" '. *Allgemeine Zeitschrift für Philosophie*, 1, 1976, pp. 34–61.